Leadership in Energy and Environmental Design

LEED® ID&C Practice Exam

Interior Design & Construction

David M. Hubka, DE, LEED AP

Professional Publications, Inc. • Belmont, California

The Benefits of Registering Your Book with PPI

- Get the latest exam news
- Hear about updates and corrections to your book
- Save with special discounts
- Succeed with exclusive exam tips and strategies

Register your book at **www.ppi2pass.com/register**.

Report Errors and View Corrections for this Book

PPI is grateful to every reader who notifies us of a possible error. Your feedback allows us to improve the quality and accuracy of our products. You can report errata and view corrections at **www.ppi2pass.com/errata**.

LEED® and USGBC® are registered trademarks of the U.S. Green Building Council. PPI® is not affiliated with the U.S. Green Building Council (USGBC) or the Green Building Certification Institute (GBCI) and does not administer the LEED AP program or LEED Green Building Rating System. PPI does not claim any endorsement or recommendation of its products or services by USGBC or GBCI.

Energy Star® (ENERGY STAR®) is a registered trademark of the U.S. Environmental Protection Agency (EPA).

LEED ID&C PRACTICE EXAM: INTERIOR DESIGN & CONSTRUCTION

Current printing of this edition: 1

Printing History

edition number	printing number	update
1	1	New book.

Copyright © 2009 by Professional Publications, Inc. (PPI). All rights reserved. No part of this publication may be reproduced, stored in a retrieval system, or transmitted, in any form or by any means, electronic, mechanical, photocopying, recording, or otherwise, without the prior written permission of the publisher.

Printed in the United States of America

PPI
1250 Fifth Avenue, Belmont, CA 94002
(650) 593-9119
www.ppi2pass.com

ISBN: 978-1-59126-187-2

Table of Contents

Preface and Acknowledgments v

Introduction vii
    About the LEED Credentialing Program vii
    About the LEED AP Interior Design & Construction Exam. vii
    Taking the LEED Credentialing Exams ix
    How to Use This Book x

References. xi
    Primary References for Part One of the LEED AP Interior Design & Construction Exam xi
    Secondary References for Part One of the LEED AP Interior Design & Construction Exam. . . xii
    References for Part Two of the LEED AP Interior Design & Construction Exam xii

Practice Exam Part One 1

Practice Exam Part Two 23

Practice Exam Part One Solutions 47

Practice Exam Part Two Solutions 67

Preface and Acknowledgments

This book, which is written to help you study for and pass the LEED AP Interior Design & Construction credentialing exam, is a product of my passion for green building. Soon after I began designing mechanical systems, I recognized that many contractors, engineers, product vendors, and architects desired a fundamental understanding of LEED. As a result, I began creating hour-long seminars discussing the impact of buildings on the environment and the growing presence of LEED for the benefit of local building professionals. The success of these local seminars paved the way for me to present LEED exam prep training seminars throughout the country. The problems in this book, which are unique, based on my experiences, and representative of those on the actual exam, are outgrowths of those presentations and seminars.

I wish to acknowledge those who have helped me create this book. First, thanks go to the incredibly talented professionals working at Total Mechanical. Their combined experience across all mechanical trades has proven to be an invaluable resource for me; without their assistance I would not have the broad knowledge of LEED that I have today. Thanks also go to Mike Hyde, Total Mechanical's Chief HVAC Engineer. He continually provides forward-thinking solutions to our LEED projects. I am deeply appreciative of the Mechanical Service Contractors of America (MSCA) for hiring me to provide my first LEED exam prep training, and of Barb Dolim, MSCA Executive Director, for providing direction as I developed the training. I would also like to thank all of my LEED seminar attendees, who continually contribute to my understanding and appreciation of the rewards and challenges of green building.

Next, thank you to those at PPI who helped in the process of creating this book, including director of new product development Sarah Hubbard, director of production Cathy Schrott, editor Courtnee Crystal, and typesetter and cover designer Amy Schwertman.

Finally, I would like to thank my wife, Dana, for her support throughout the entire process.

Despite all the help and support I had, any mistakes you find are mine. If you find any mistakes, please report them through **www.ppi2pass.com/errata**. Corrections will be posted on the PPI website and incorporated into this book when it is reprinted.

Good luck on the exam and in all your green building efforts.

David M. Hubka, DE, LEED AP

Introduction

About the LEED Credentialing Program

The Green Building Certification Institute (GBCI) offers credentialing opportunities to professionals who demonstrate knowledge of Leadership in Energy and Environmental Design (LEED) green building practices. *LEED ID&C Practice Exams: Interior Design & Construction* prepares you for both parts of the LEED AP Interior Design & Construction exam.

GBCI's LEED credentialing program has three tiers. The first tier corresponds to the LEED Green Associate exam. According to the *LEED Green Associate Candidate Handbook*, this exam confirms that you have the knowledge and skills necessary to understand and support green design, construction, and operations. When you pass the LEED Green Associate exam, you will earn the LEED Green Associate credential.

The second tier, which corresponds to the LEED AP specialty exams, confirms your deeper and more specialized knowledge of green building practices. GBCI currently has planned five tracks for the LEED AP exams: LEED AP Homes, LEED AP Operations & Maintenance, LEED AP Building Design & Construction, LEED AP Interior Design & Construction, and LEED AP Neighborhood Development. The LEED AP exams are based on the corresponding LEED reference guide and rating systems and other references. When you pass the LEED Green Associate exam along with any LEED AP specialty exam, you will earn the LEED AP credential.

The third tier, called LEED AP Fellow, will distinguish professionals with an exceptional depth of knowledge, experience, and accomplishments with LEED green building practices. This distinction will be attainable through extensive LEED project experience, not by taking an exam.

For more information about LEED credentialing, visit **www.ppi2pass.com/LEEDhome**.

About the LEED AP Interior Design & Construction Exam

The LEED AP Interior Design & Construction exam (and the practice exam in this book) contains 200 problems. The first part of the exam (whose specifications are identical to those of the LEED Green Associate exam) contains 100 questions that test your knowledge of green building practices and principles, as well as your familiarity with LEED requirements, resources, and processes. Accordingly, GBCI categorizes the exam questions into the following seven subject areas.

- *Synergistic Opportunities and LEED Application Process* (project requirements; costs; green resources; standards that support LEED credit; credit interactions; Credit Interpretation Requests and rulings; exemplary performance credits; components of LEED Online and project registration; components of LEED Scorecard; components of letter templates; strategies to achieve credit; project boundary; LEED boundary; property boundary; prerequisites and/or minimum program requirements for LEED certification; preliminary rating; multiple certifications for same building; occupancy requirements; USGBC policies; requirements to earn LEED AP credit)
- *Project Site Factors* (community connectivity: transportation and pedestrian access; zoning requirements; heat islands)
- *Water Management* (types and quality of water; water management)
- *Project Systems and Energy Impacts* (environmental concerns; green power)
- *Acquisition, Installation, and Management of Project Materials* (recycled materials; regionally harvested and manufactured materials; construction waste management)
- *Stakeholder Involvement in Innovation* (integrated project team criteria; durability planning and management; innovative and regional design)
- *Project Surroundings and Public Outreach* (codes)

The second part of the exam contains an additional 100 questions that test your knowledge of subject areas unique to the interior design and construction of a building. These are the subject areas covered in the second part of this book's practice exam. GBCI has identified these as follows.

- *Project Site Factors* (community connectivity and services; lighting)
- *Water Management* (water treatment; stormwater; specialized equipment needs)
- *Project Systems and Energy Impacts* (energy performance policies; building components; on-site renewable energy; requirements for third-party relationships; alternate rating systems; energy performance measurement; energy trade-offs; energy usage; specialized equipment power needs)
- *Acquisition, Installation, and Management of Project Materials* (building reuse; rapidly renewable materials; acquisition of materials; chemical management policy and audit; environmental management plan)
- *Improvements to the Indoor Environment* (minimum ventilation requirement; tobacco smoke control; air quality; ventilation effectiveness; indoor air quality: pre-construction, during construction, before occupancy, and during occupancy; low-emitting materials; indoor/outdoor chemical and pollutant control; lighting controls; thermal controls; views; ergonomics; acoustics; types of building spaces)
- *Stakeholder Involvement in Innovation* (design workshop/charrette; ways to earn credit; education of homeowner or tenant; education of building manager)
- *Project Surroundings and Public Outreach* (information on available community resources; reduced parking methods; transit-oriented development; ADA/universal access)

Taking the LEED Credentialing Exams

To apply for a LEED credentialing exam, you must agree to the disciplinary policy and credential maintenance requirements and submit to an application audit. To be eligible to take the LEED Green Associate exam, one of the following must be true.

- Your line of work is in a sustainable field.
- You have documented experience supporting a LEED-registered project.
- You have attended an education program that addresses green building principles.

To be eligible to take a LEED AP exam, you must have documented experience with a LEED-registered project within the three years prior to your application submittal.

The LEED credentialing exams are administered by computer at Prometric test sites. Prometric is a third-party testing agency with over 250 testing locations in the United States and hundreds of centers globally. To schedule an exam, you must first apply at www.gbci.org to receive an eligibility ID number. Then, you must go to the Prometric website at www.prometric.com/gbci to schedule and pay for the exam. If you need to reschedule or cancel your exam, you must do so directly through Prometric.

The LEED credentialing exam questions are multiple choice with four or more answer options for each question. If more than one option must be selected to correctly answer a question, the question stem will indicate how many options you must choose. Each 100-question exam lasts two hours, giving you a bit more than one and a half minutes per question. The bulk of the questions are non-numerical. Calculators are not allowed or provided, but only basic math is needed to correctly solve any quantitative questions. No reference materials or other supplies may be brought into the exam room, though a pencil and scratch paper will be provided by the testing center. (References are not provided.) The only thing you need to bring with you on exam day is your identification.

Your testing experience begins with an optional brief tutorial to introduce you to the testing computer's functions. When you've finished the tutorial, questions and answer options are shown on a computer screen, and the computer keeps track of which options you choose. Because points are not deducted for incorrectly answered questions, you should mark an answer to every question. For answers you are unsure of, make your best guess and flag the question for later review. If you decide on a different answer later, you can change it, but if you run out of time before getting to all your flagged questions, you still will have given a response to each one. Be sure to mark the correct number of options for each question. There is no partial credit for incomplete answers (or for selecting only some of the correct options).

If you are taking both the first tier (LEED Green Associate) and the second tier (LEED AP) exams on the same day, at the end of your first session the computer will ask you if you are ready to take the second tier. You can take a short break at this time. The second tier's two hours begins when you click "yes" to indicate that you are ready.

To ensure that all candidates' chances of passing remain constant regardless of the difficulty of the specific questions administered on any given exam, GBCI converts the raw exam score to a scaled score, with the total number of points set at 200 and a minimum passing score of 170. In this way, you are not penalized if the exam taken is more difficult than another exam. Instead, in such a case, fewer questions must be answered correctly to achieve a passing score. Your scaled score (or scores, if you are taking both tiers on the same day) is reported on the screen upon completing the exam. A brief optional exit survey completes the exam experience.

When you pass the LEED Green Associate exam, a LEED Green Associate certificate will be sent to you in the mail. If you take and pass both exams, a LEED AP certificate will be sent to you in the mail. If you take both exams but pass only the LEED AP exam, you will need to register again and retake and pass the LEED Green Associate exam before you receive any LEED credential.

How to Use This Book

There are a few ways you can use this book's practice exam. You can do an untimed review of the questions and answers to familiarize yourself with the exam format and content and determine which subjects you are weak in, using this information as a guide for studying. This book's companion volume, *LEED Prep ID&C: What You Really Need to Know to Pass the LEED AP Interior Design & Construction Exam*, will give you a complete, concise review of the subjects covered on the exam. *LEED ID&C Flashcards*, also published by PPI, will reinforce your ability to retain and recall what you've studied.

Or you can use this book to simulate the exam experience, either as a pretest before you begin your study or when you think you are fully prepared. In that case, treat this practice exam as though it were the real thing. Don't look at the questions or answers ahead of time. Put away your study materials and references, set a timer for two hours, and solve as many questions as you can within the time limit. Practice exam-like time management. Fill in the provided bubble sheet with your best guess on every question regardless of your certainty and mark the answers to revisit if time permits. If you finish before the time is up, review your work. If you are unable to finish within the time limit, make a note of where you were after two hours, but continue on to complete the exam. Keep track of your time to see how much faster you will need to work to finish the actual exam within two hours.

After taking the practice exam, check your answers against the answer key. Consider a problem correctly answered only if you have selected all of the required options (and no others). Calculate the percentage correct. Though the actual exam score will be scaled, aim for getting at least 70% (70 questions) of the practice exam's questions correct. The fully explained solutions are a learning tool. In addition to reading the solutions to the questions you answered incorrectly, read the explanations to those you answered correctly. Categorize your incorrect responses by exam subject to help you determine the areas you need to study. Use GBCI's list of references and PPI's *LEED Prep ID&C* to guide your preparation. Though this exam reflects the breadth and depth of the content on the actual exam, use your best judgment when determining the subjects you need to review.

References

The LEED ID&C Practice Exam: Interior Design & Construction is based on the following references, identified by the Green Building Certification Institute (GBCI) in its *LEED AP Interior Design & Construction Candidate Handbook*. Most of these references are available electronically. You can find links to these references on PPI's website, **www.ppi2pass.com/ LEEDreferences**.

Primary References for Part One of the LEED AP Interior Design & Construction Exam

Bernheim, Anthony, and William Reed. "Part II: Pre-Design Issues." *Sustainable Building Technical Manual*. Public Technology, Inc. 1996.

Cost of Green Revisited: Reexamining the Feasibility and Cost Impact of Sustainable Design in Light of Increased Market Adoption. Sacramento, CA: Davis Langdon, 2007.

Guidance on Innovation & Design (ID) Credits. Announcement. Washington, DC: U.S. Green Building Council, 2004.

Guidelines for CIR Customers. Announcement. Washington, DC: U.S. Green Building Council, 2007.

LEED for Interior Design & Construction Reference Guide. Washington, DC: U.S. Green Building Council, 2009.

LEED Technical and Scientific Advisory Committee. *The Treatment by LEED of the Environmental Impact of HVAC Refrigerants*. Washington, DC: U.S. Green Building Council, 2004.

Secondary References for Part One of the LEED AP Interior Design & Construction Exam

AIA Integrated Project Delivery: A Guide. American Institute of Architects, 2007.

Americans with Disabilities Act: Standards for Accessible Design. 28 CFR Part 36. Washington, DC: Code of Federal Regulations, 1994.

"Codes and Standards." Washington, DC: International Code Council, 2009.

"Construction and Building Inspectors." *Occupational Outlook Handbook.* Washington, DC: Bureau of Labor Statistics, 2009.

Frankel, Mark, and Cathy Turner. *Energy Performance of LEED for New Construction Buildings: Final Report.* Vancouver, WA: New Buildings Institute and U.S. Green Building Council, 2008.

GSA 2003 Facilities Standards. Washington, DC: General Services Administration, 2003.

Guide to Purchasing Green Power: Renewable Electricity, Renewable Energy Certifications, and On-Site Renewable Generation. Washington, DC: Environmental Protection Agency, 2004.

Kareis, Brian. *Review of ANSI/ASHRAE Standard 62.1-2004: Ventilation for Acceptable Indoor Air Quality.* Greensboro, NC: Workplace Group, 2007.

Lee, Kun-Mo, and Haruo Uehara. *Best Practices of ISO 14021: Self-Declared Environmental Claims.* Suwon, Korea: Ajou University, 2003.

LEED Steering Committee. *Foundations of the Leadership in Energy and Environmental Design Environmental Rating System: A Tool for Market Transformation.* Washington, DC: U.S. Green Building Council, 2006.

References for Part Two of the LEED AP Interior Design & Construction Exam

Bernheim, Anthony, and William Reed. "Part II: Pre-Design Issues." *Sustainable Building Technical Manual.* Public Technology, Inc. 1996.

Cost of Green Revisited: Reexamining the Feasibility and Cost Impact of Sustainable Design in Light of Increased Market Adoption. Sacramento, CA: Davis Langdon, 2007.

Guidance on Innovation & Design (ID) Credits. Announcement. Washington, DC: United States Green Building Council, 2004.

Guidelines for CIR Customers. Announcement. Washington, DC: U.S. Green Building Council, 2007.

LEED Online Sample Credit Templates. Washington, DC: United States Green Building Council, 2009.

LEED for Interior Design & Construction Reference Guide. Washington, DC: United States Green Building Council, 2009.

Practice Exam Part One

1. Ⓐ Ⓑ Ⓒ Ⓓ
2. Ⓐ Ⓑ Ⓒ Ⓓ Ⓔ
3. Ⓐ Ⓑ Ⓒ Ⓓ
4. Ⓐ Ⓑ Ⓒ Ⓓ Ⓔ
5. Ⓐ Ⓑ Ⓒ Ⓓ Ⓔ
6. Ⓐ Ⓑ Ⓒ Ⓓ Ⓔ
7. Ⓐ Ⓑ Ⓒ Ⓓ Ⓔ
8. Ⓐ Ⓑ Ⓒ Ⓓ Ⓔ
9. Ⓐ Ⓑ Ⓒ Ⓓ Ⓔ
10. Ⓐ Ⓑ Ⓒ Ⓓ Ⓔ
11. Ⓐ Ⓑ Ⓒ Ⓓ
12. Ⓐ Ⓑ Ⓒ Ⓓ
13. Ⓐ Ⓑ Ⓒ Ⓓ
14. Ⓐ Ⓑ Ⓒ Ⓓ Ⓔ
15. Ⓐ Ⓑ Ⓒ Ⓓ Ⓔ
16. Ⓐ Ⓑ Ⓒ Ⓓ Ⓔ
17. Ⓐ Ⓑ Ⓒ Ⓓ
18. Ⓐ Ⓑ Ⓒ Ⓓ
19. Ⓐ Ⓑ Ⓒ Ⓓ
20. Ⓐ Ⓑ Ⓒ Ⓓ Ⓔ
21. Ⓐ Ⓑ Ⓒ Ⓓ Ⓔ
22. Ⓐ Ⓑ Ⓒ Ⓓ Ⓔ
23. Ⓐ Ⓑ Ⓒ Ⓓ Ⓔ
24. Ⓐ Ⓑ Ⓒ Ⓓ
25. Ⓐ Ⓑ Ⓒ Ⓓ Ⓔ

26. Ⓐ Ⓑ Ⓒ Ⓓ Ⓔ
27. Ⓐ Ⓑ Ⓒ Ⓓ
28. Ⓐ Ⓑ Ⓒ Ⓓ Ⓔ
29. Ⓐ Ⓑ Ⓒ Ⓓ
30. Ⓐ Ⓑ Ⓒ Ⓓ
31. Ⓐ Ⓑ Ⓒ Ⓓ Ⓔ
32. Ⓐ Ⓑ Ⓒ Ⓓ
33. Ⓐ Ⓑ Ⓒ Ⓓ
34. Ⓐ Ⓑ Ⓒ Ⓓ Ⓔ
35. Ⓐ Ⓑ Ⓒ Ⓓ
36. Ⓐ Ⓑ Ⓒ Ⓓ Ⓔ Ⓕ
37. Ⓐ Ⓑ Ⓒ Ⓓ
38. Ⓐ Ⓑ Ⓒ Ⓓ Ⓔ
39. Ⓐ Ⓑ Ⓒ Ⓓ
40. Ⓐ Ⓑ Ⓒ Ⓓ Ⓔ
41. Ⓐ Ⓑ Ⓒ Ⓓ
42. Ⓐ Ⓑ Ⓒ Ⓓ Ⓔ Ⓕ
43. Ⓐ Ⓑ Ⓒ Ⓓ Ⓔ
44. Ⓐ Ⓑ Ⓒ Ⓓ Ⓔ
45. Ⓐ Ⓑ Ⓒ Ⓓ
46. Ⓐ Ⓑ Ⓒ Ⓓ
47. Ⓐ Ⓑ Ⓒ Ⓓ
48. Ⓐ Ⓑ Ⓒ Ⓓ Ⓔ
49. Ⓐ Ⓑ Ⓒ Ⓓ
50. Ⓐ Ⓑ Ⓒ Ⓓ

51. Ⓐ Ⓑ Ⓒ Ⓓ Ⓔ
52. Ⓐ Ⓑ Ⓒ Ⓓ Ⓔ Ⓕ
53. Ⓐ Ⓑ Ⓒ Ⓓ
54. Ⓐ Ⓑ Ⓒ Ⓓ Ⓔ
55. Ⓐ Ⓑ Ⓒ Ⓓ Ⓔ
56. Ⓐ Ⓑ Ⓒ Ⓓ Ⓔ
57. Ⓐ Ⓑ Ⓒ Ⓓ Ⓔ Ⓕ Ⓖ
58. Ⓐ Ⓑ Ⓒ Ⓓ Ⓔ
59. Ⓐ Ⓑ Ⓒ Ⓓ
60. Ⓐ Ⓑ Ⓒ Ⓓ Ⓔ
61. Ⓐ Ⓑ Ⓒ Ⓓ Ⓔ
62. Ⓐ Ⓑ Ⓒ Ⓓ
63. Ⓐ Ⓑ Ⓒ Ⓓ Ⓔ
64. Ⓐ Ⓑ Ⓒ Ⓓ
65. Ⓐ Ⓑ Ⓒ Ⓓ
66. Ⓐ Ⓑ Ⓒ Ⓓ Ⓔ Ⓕ
67. Ⓐ Ⓑ Ⓒ Ⓓ Ⓔ
68. Ⓐ Ⓑ Ⓒ Ⓓ
69. Ⓐ Ⓑ Ⓒ Ⓓ
70. Ⓐ Ⓑ Ⓒ Ⓓ
71. Ⓐ Ⓑ Ⓒ Ⓓ Ⓔ
72. Ⓐ Ⓑ Ⓒ Ⓓ
73. Ⓐ Ⓑ Ⓒ Ⓓ
74. Ⓐ Ⓑ Ⓒ Ⓓ
75. Ⓐ Ⓑ Ⓒ Ⓓ

76. Ⓐ Ⓑ Ⓒ Ⓓ Ⓔ
77. Ⓐ Ⓑ Ⓒ Ⓓ Ⓔ
78. Ⓐ Ⓑ Ⓒ Ⓓ
79. Ⓐ Ⓑ Ⓒ Ⓓ
80. Ⓐ Ⓑ Ⓒ Ⓓ
81. Ⓐ Ⓑ Ⓒ Ⓓ Ⓔ
82. Ⓐ Ⓑ Ⓒ Ⓓ
83. Ⓐ Ⓑ Ⓒ Ⓓ
84. Ⓐ Ⓑ Ⓒ Ⓓ Ⓔ Ⓕ
85. Ⓐ Ⓑ Ⓒ Ⓓ
86. Ⓐ Ⓑ Ⓒ Ⓓ Ⓔ Ⓕ
87. Ⓐ Ⓑ Ⓒ Ⓓ
88. Ⓐ Ⓑ Ⓒ Ⓓ Ⓔ
89. Ⓐ Ⓑ Ⓒ Ⓓ Ⓔ

90. Ⓐ Ⓑ Ⓒ Ⓓ Ⓔ
91. Ⓐ Ⓑ Ⓒ Ⓓ
92. Ⓐ Ⓑ Ⓒ Ⓓ Ⓔ
93. Ⓐ Ⓑ Ⓒ Ⓓ Ⓔ
94. Ⓐ Ⓑ Ⓒ Ⓓ
95. Ⓐ Ⓑ Ⓒ Ⓓ
96. Ⓐ Ⓑ Ⓒ Ⓓ Ⓔ Ⓕ Ⓖ
97. Ⓐ Ⓑ Ⓒ Ⓓ Ⓔ
98. Ⓐ Ⓑ Ⓒ Ⓓ
99. Ⓐ Ⓑ Ⓒ Ⓓ Ⓔ
100. Ⓐ Ⓑ Ⓒ Ⓓ

1. Who can view CIRs posted to the USGBC website? (Choose two.)
    (A) individuals with a USGBC website account
    (B) registered employees of USGBC member companies
    (C) LEED Accredited Professionals
    (D) registered project team members

2. LEED project teams can earn an "extra" point by achieving _____ performance. (Choose two.)
    (A) exemplary
    (B) ideal
    (C) innovative
    (D) original
    (E) perfect

3. Potable water can also be called _____.
    (A) drinking water
    (B) graywater
    (C) blackwater
    (D) rainwater

4. To increase a LEED project's chances of success, team members should be involved with which of the following project phases? (Choose three.)
    (A) concept
    (B) design development
    (C) inspector site visits
    (D) ongoing commissioning
    (E) construction

5. LEED generally groups credits by credit categories, but the LEED O&M reference guide introduction describes an alternative way of grouping the credits: by their functional characteristics. Which of the following are identified as functional characteristic groups? (Choose two.)
    (A) Administration
    (B) Materials In
    (C) Sustainable Sites
    (D) Waste Management
    (E) Water Efficiency

6. A LEED project is more likely to stay within budget when the team does which of the following? (Choose three.)

    (A) submits documentation for as few LEED credits as possible
    (B) adheres to the plan throughout project
    (C) aligns goals with budget
    (D) contacts USGBC for budget guidance
    (E) establishes project goals and expectations

7. Projects located in urban areas can utilize which of the following to meet LEED open space requirements? (Choose two.)

    (A) accessible roof decks
    (B) landscaping with indigenous plants
    (C) non-vehicular, pedestrian-orientated hardscapes
    (D) on-site photovoltaics
    (E) pervious parking lots

8. Which of the following can affect a building's energy efficiency? (Choose three.)

    (A) building orientation
    (B) envelope thermal efficiency
    (C) HVAC system sizing
    (D) refrigerant selection
    (E) VOC content of building materials

9. Which of the following are included in calculations used to determine life-cycle costs? (Choose two.)

    (A) equipment
    (B) facility alterations
    (C) maintenance
    (D) occupant transportation
    (E) utilities

10. Which of the following are NOT common benefits of daylighting? (Choose two.)

    (A) increased productivity
    (B) reduced air pollution
    (C) reduced heat island effect
    (D) reduced light pollution
    (E) reduced operating costs

11. The LEED CI rating system can be applied to which of the following projects? (Choose two.)

    (A) major envelope renovation of a building
    (B) renovation of part of an owner-occupied building
    (C) tenant infill of an existing building
    (D) upgrades to the operation and maintenance of an existing facility

12. Certain prerequisites and credits require project teams to create policies. What information should be included in a policy model? (Choose two.)

    (A) performance period
    (B) policy author
    (C) responsible party
    (D) time period

13. The primary function of which of the following is to encourage sustainable building design, construction, and operation?

    (A) chain of custody
    (B) the LEED rating systems
    (C) standard operating procedures
    (D) waste reduction program

14. What information is required to register a project for LEED certification? (Choose three.)

    (A) list of LEED project team members
    (B) name of LEED AP who will be working on project
    (C) primary contact information
    (D) project owner information
    (E) project type

15. Which of the following are among the five basic steps for pursing LEED for Homes certification? (Choose two.)

    (A) achieve certification as a LEED home
    (B) become a USGBC member company
    (C) create list of LEED certified products to be used
    (D) include appliances only if they are Energy Star-rated
    (E) market and sell the home

16. Fire suppression systems that use which of the following will contribute the least to ozone depletion? (Choose two.)
    (A) chlorofluorocarbons
    (B) halons
    (C) hydrochlorofluorocarbons
    (D) hydrofluorocarbons
    (E) water

17. Which of the following describes the *property area*?
    (A) area of the project site impacted by the building and hardscapes
    (B) area of the project site impacted by hardscapes only
    (C) area of the project site impacted by the building only
    (D) total area of the site including constructed and non-constructed areas

18. Which LEED rating system includes performance periods and requires recertification to maintain the building's LEED certification?
    (A) LEED CI
    (B) LEED CS
    (C) LEED EBO&M
    (D) LEED NC

19. For a credit uploaded to LEED Online, a white check mark next to a credit name indicates that the credit is _____.
    (A) being pursued and no online documentation has been uploaded
    (B) being pursued and some online documentation has been uploaded
    (C) complete and ready for submission
    (D) not being pursued

20. Remodeling an older existing building may help a project team achieve credit for building reuse and prevent a project team from achieving credit for _____. (Choose three.)
    (A) controllability of systems
    (B) energy use reduction
    (C) regional materials
    (D) sustainable site use
    (E) water use reduction

21. LEED CS (rather than LEED NC) should be pursued when which of the following items are outside the control of the building owner? (Choose three.)
    (A) envelope insulation
    (B) interior finishes
    (C) lighting
    (D) mechanical distribution
    (E) site selection

22. LEED Online provides a means for _____. (Choose two.)
    (A) code officials to access project documentation
    (B) product vendors to advertize
    (C) project team members to analyze anticipated building energy performance
    (D) project administrators to manage LEED projects
    (E) project team members to manage LEED prerequisites and credits

23. Installing a green roof can help a project team achieve which of the following LEED credits? (Choose two.)
    (A) Development Density and Community Connectivity
    (B) Heat Island Effect
    (C) Light Pollution Reduction
    (D) Site Selection
    (E) Stormwater Design

24. Using light bulbs with low mercury content, long life, and high lumen output will result in which of the following?
    (A) improved indoor air quality
    (B) increased light pollution
    (C) reduced light pollution
    (D) reduced toxic waste

25. Which of the following are covered in the Sustainable Sites credit category? (Choose three.)
    (A) light to night sky
    (B) light trespass
    (C) on-site renewable energy
    (D) stormwater mitigation
    (E) refrigerants

26. Which of the following tasks must be part of a durability plan? (Choose two.)
    - (A) assign responsibilities for plan implementation
    - (B) evaluate durability risks of project
    - (C) incorporate durability strategies into design
    - (D) research local codes
    - (E) submit CIR for available list of durability strategies

27. What is the role of the TSAC? (Choose two.)
    - (A) ensure the technical soundness of the LEED reference guides and training
    - (B) maintain technical rigor and consistency in the development of LEED credits
    - (C) resolve issues to maintain consistency across different LEED rating systems
    - (D) respond to CIRs submitted by LEED project teams

28. Project teams can earn credit for purchasing sustainable ongoing consumables within the LEED EBO&M rating system. Ongoing consumables containing a defined amount of _____ are considered sustainable. (Choose three.)
    - (A) material certified by the Rainforest Alliance
    - (B) preindustrial material
    - (C) rapidly renewable material
    - (D) regionally extracted material
    - (E) salvaged material

29. Within an existing building undergoing major renovations, a tenant is pursuing LEED certification. Which rating system should the tenant use to earn a LEED plaque?
    - (A) LEED CI
    - (B) LEED CS
    - (C) LEED EBO&M
    - (D) LEED NC

30. Regional Priority credits vary depending on a project's _____.
    - (A) certification level
    - (B) community connectivity
    - (C) development density
    - (D) geographic location

31. Which of the following costs should be considered prior to pursuing a LEED credit? (Choose three.)
    - (A) application review cost
    - (B) construction cost
    - (C) documentation cost
    - (D) registration cost
    - (E) soft cost

32. Which rating system includes CIRs, appeals, and performance periods?

(A) LEED CI
(B) LEED CS
(C) LEED EBO&M
(D) LEED NC

33. Which standard addresses energy-efficient building design?

(A) ASHRAE 55-2004
(B) ANSI/ASHRAE 52.2-1999
(C) ANSI/ASHRAE 62.1-2007
(D) ANSI/ASHRAE/IESNA 90.1-2007

34. Which of the following credits will NOT be directly affected by a project team intending to increase a building's ventilation? (Choose two.)

(A) Controllability of Systems
(B) Enhanced Commissioning
(C) Indoor Chemical and Pollutant Source Control
(D) Measurement and Verification
(E) Optimize Energy Performance

35. Which of the following describes the purpose of a chain-of-custody document?

(A) track the movement of products from extraction to the production site
(B) track the movement of wood products from the forest to the building site
(C) verify rapidly renewable materials
(D) verify recycled content

36. A material must be which of the following for it to qualify as a *regional material* within the LEED rating systems? (Choose two.)

(A) FSC-certified
(B) made from 10% post-consumer content
(C) made from products that take 10 years or less to grow
(D) manufactured within 500 miles of the project site
(E) permanently installed on the project site
(F) used to create process equipment

37. A project team's choice of paint will NOT affect which of the following credits?

(A) Heat Island Effect
(B) Low-Emitting Materials
(C) Materials Reuse
(D) Rapidly Renewable Materials

38. Which of the following do all LEED rating systems contain? (Choose three.)

    (A) core credits
    (B) educational credits
    (C) innovation credits
    (D) operational credits
    (E) prerequisites

39. Which of the following is NOT an example of a durable good?

    (A) computer
    (B) door
    (C) landscaping equipment
    (D) office desk

40. Which of the following would most likely be affected by an increase in ventilation? (Choose two.)

    (A) construction costs
    (B) interior temperature set points
    (C) local climate
    (D) operational costs
    (E) refrigerant management

41. Which items should be considered when selecting refrigerants for a building's HVAC & R system? (Choose two.)

    (A) global depletion potential
    (B) global warming potential
    (C) ozone depletion potential
    (D) ozone warming potential

42. Which information is required to set up a personal account on the USGBC website? (Choose three.)

    (A) company name
    (B) email address
    (C) industry sector
    (D) LEED credentialing exam date
    (E) phone number
    (F) prior LEED projects worked on

43. Project teams must comply with which of the following CIRs? (Choose two.)

    (A) CIRs appealed prior to project registration, for all rating systems
    (B) CIRs reviewed by a TAG for their own project
    (C) CIRs reviewed by a TAG prior to project registration, for projects within the project's climate region
    (D) CIRs posted prior to project application, for the applicable rating system only
    (E) CIRs posted prior to project completion, for projects within their own project's climate region

44. Which of the following describe methods of earning exemplary performance credit? (Choose two.)

    (A) achieve 75% of the credits in each LEED category
    (B) achieve every LEED credit in the Energy and Atmosphere category
    (C) achieve the next incremental level of an existing credit
    (D) double the requirements of an existing credit
    (E) pursue LEED CI certification within a LEED CS-certified building

45. Which of the following potable water conserving strategies may help a project team achieve a LEED point? (Choose two.)

    (A) collecting blackwater for landscape irrigation
    (B) collecting rainwater for sewage conveyance
    (C) installing an on-site septic tank
    (D) using cooling condensate for cooling tower make-up

46. Which of the following requires an explanation of the proposed credit requirements?

    (A) CIR submittal
    (B) Innovation in Design exemplary performance submittal
    (C) LEED credit equivalence submittal
    (D) LEED Online letter template submittal

47. Which of the following organizations defines the off-site renewable energy sources eligible for LEED credits?

    (A) Center for Research and Development of Green Power
    (B) Center for Resource Solutions
    (C) Department of Energy
    (D) Energy Star

48. The International Code Council (ICC) includes which of the following codes? (Choose three.)

    (A) International Building Automation Code (IBAC)
    (B) International Energy Conservation Code (IECC)
    (C) International Lighting Code (ILC)
    (D) International Mechanical Code (IMC)
    (E) International Plumbing Code (IPC)

49. Which of the following is true?

    (A) A LEED project administrator must be a LEED AP.
    (B) Buildings can be LEED accredited.
    (C) Companies can be USGBC members.
    (D) People can be LEED certified.

50. In addition to satisfying all prerequisites, what is the minimum percentage of points that a project team can earn to achieve LEED Platinum certification?

    (A) 70%
    (B) 80%
    (C) 90%
    (D) 100%

51. LEED submittal templates provide which of the following? (Choose two.)

    (A) a list of potential strategies to achieve a credit or prerequisite
    (B) a means to modify project documentation
    (C) a means to submit a project for review
    (D) a means to review and submit CIRs
    (E) the project's final scorecard

52. According to the *Sustainable Building Technical Manual*, which of the following are key steps of an environmentally responsive design process? (Choose three.)

    (A) bid
    (B) design
    (C) post-design
    (D) pre-design
    (E) rebid
    (F) vendor selection

53. Which is true of the LEED CI rating system?
    - (A) Precertification allows the owner to market to potential tenants.
    - (B) Projects can earn a point for prohibiting smoking within the tenant space.
    - (C) Projects can earn half points under the Site Selection credit.
    - (D) Projects must recertify every five years to maintain certification status.

54. Once a project undergoes a construction review, what are the potential rulings for each submitted prerequisite and credit? (Choose three.)
    - (A) anticipated
    - (B) clarify
    - (C) deferred
    - (D) denied
    - (E) earned

55. What is the role of the LEED Steering Committee? (Choose two.)
    - (A) delegate responsibility and oversee all LEED committee activities
    - (B) develop LEED accreditation exams
    - (C) ensure that LEED and its supporting documentation is technically sound
    - (D) establish and enforce LEED direction and policy
    - (E) respond to CIRs submitted by LEED project teams

56. The LEED Online workspace allows the LEED project administrator to do which of the following? (Choose three.)
    - (A) apply for LEED EBO&M precertification
    - (B) assign credits to project team members
    - (C) build a project team
    - (D) review credit appeals submitted by other project teams
    - (E) submit projects for review

57. Which of the following items are free? (Choose three.)
    - (A) a project's first CIR
    - (B) LEED BD & C reference guide
    - (C) LEED brochure
    - (D) LEED certification
    - (E) LEED for Homes rating system
    - (F) sample LEED submittal templates
    - (G) usgbc.org account

58. What happens if a CIR extends beyond the expertise of the assigned TAG? (Choose two.)
    - (A) additional response time may be incurred
    - (B) it is rejected
    - (C) it is sent to the LEED Steering Committee
    - (D) it must be submitted as an Innovation in Design credit
    - (E) the TAG will provide the ruling

59. What becomes available once a project is registered?
    - (A) LEED project tools
    - (B) posted CIRs
    - (C) posted credit appeals
    - (D) sample submittal templates

60. The LEED rating systems require compliance with which of the following? (Choose two.)
    - (A) codes and regulations that address asbestos and water discharge
    - (B) codes and regulations that address PCBs and water management
    - (C) referenced standards that address fixture performance requirements for water use
    - (D) referenced standards that address sustainable forest management practices
    - (E) referenced standards that address VOCs

61. Which of the following steps are part of creating an integrated project team? (Choose two.)
    - (A) include members from varying industry sectors
    - (B) include product vendors in the design phase
    - (C) involve a commissioning authority in team members' selection
    - (D) involve the LEED AP in design integration
    - (E) involve the team in different project phases

62. Facilities undergoing minor alterations and system upgrades must follow which rating system?
    - (A) LEED CI
    - (B) LEED CS
    - (C) LEED EBO&M
    - (D) LEED NC

63. Which of the following strategies may help a project team achieve a LEED credit? (Choose two.)

   (A) establish an erosion and sedimentation plan
   (B) establish a location for the storage and collection of recyclables
   (C) install HCFC-based HVAC & R equipment
   (D) use rainwater for sewage conveyance or landscape irrigation
   (E) remediate contaminated soil

64. Which of the following statements is true?

   (A) CIRs are reviewed by a TAG at no additional charge once a project is registered.
   (B) The first step toward LEED certification is passing the Green Associate exam.
   (C) The LEED rating system only applies to commercial buildings.
   (D) To achieve LEED certification, every prerequisite and a minimum number of credits must be achieved.

65. Installing ground source heat pumps would help a project team achieve which of the following credits or prerequisites?

   (A) Green Power
   (B) Optimize Energy Performance
   (C) On-Site Renewable Energy
   (D) Water Use Reduction

66. What are the goals of the Portfolio Program? (Choose two.)

   (A) create a volume accreditation path
   (B) encourage global adoption of sustainable green building practices
   (C) exceed the requirements of ANSI/ASHRAE/IESNA 90.1-2007 by a rating system-designated percentage
   (D) offer a volume certification path
   (E) provide a streamlined certification process for large-scale projects
   (F) provide a template of key data for the design team members to compile

67. When should a LEED project's budget be addressed? (Choose two.)

   (A) before the design phase
   (B) during the construction phase
   (C) during the selection of construction team
   (D) during the selection of design team
   (E) upon project completion

68. Which of the following could help minimize potable water used for the site landscaping and contribute toward earning a LEED landscaping credit?

    (A) designing the site or the building's roof with no landscaping
    (B) installing invasive plants
    (C) installing native or adapted plants
    (D) installing turf grass

69. How many Regional Priority points are available for LEED projects?

    (A) 2 points
    (B) 4 points
    (C) 6 points
    (D) 8 points

70. A project team chooses to group credits by functional characteristics. Credits focusing on the measurement of a building's energy performance and ozone protection would be grouped into which of the following categories?

    (A) Energy and Atmosphere
    (B) Energy Metrics
    (C) Materials Out
    (D) Site Management

71. Which of the following are considered principle durability risks? (Choose three.)

    (A) heat islands
    (B) interior moisture loads
    (C) ozone depletion
    (D) pests
    (E) ultraviolet radiation

72. What is the first step of the LEED for Homes certification process?

    (A) become a LEED AP
    (B) contact a LEED for Homes provider
    (C) register with GBCI
    (D) submit a CIR

73. HVAC & R equipment with _____ will contribute to ozone depletion and global warming.

    (A) a manufacture date prior to 2005
    (B) a relatively long equipment life
    (C) minimal refrigerant charge
    (D) refrigerant leakage

74. Sealing ventilation ducts, installing rodent- and corrosion-proof screens, and using air-sealing pump covers are strategies that could be a part of which of the following?

    (A) design charrette
    (B) durability plan
    (C) landscape management plan
    (D) PE exemption form

75. Which standard addresses thermal comfort of building occupants?

    (A) ASHRAE 55-2004
    (B) ANSI/ASHRAE 52.2-1999
    (C) ANSI/ASHRAE 62.1-2007
    (D) ANSI/ASHRAE/IESNA 90.1-2007

76. A LEED project's primary contact must submit which of the following when registering a project? (Choose three.)

    (A) email address
    (B) LEED AP certificate
    (C) LEED project history
    (D) organization name
    (E) individual's title

77. Prior to registering a LEED project, which of the following must be confirmed? (Choose two.)

    (A) necessary CIRs
    (B) precertification
    (C) project cost
    (D) project summary
    (E) project team members

78. Water lost through plant transpiration and evaporation from soil is described by the term _____.

    (A) evapotranspiration
    (B) infiltration
    (C) sublimation
    (D) surface runoff

79. Individuals with a LEED reference guide electronic access code can do which of the following?

    (A) join a LEED project as a team member
    (B) print the LEED reference guide
    (C) purchase a LEED reference guide
    (D) view a protected electronic version of the LEED reference guide

80. LEED credits and prerequisites are presented in a common format in all versions of LEED rating systems. The structure includes which of the following?

    (A) economic impact
    (B) greening opportunities
    (C) intent
    (D) submittal requirements

81. The Materials and Resources category directly addresses which of the following? (Choose two.)

    (A) habitat conservation
    (B) durable goods
    (C) energy consumption
    (D) landscape
    (E) waste stream

82. Building on which of the following sites will most likely have the smallest impact on the environment?

    (A) greenfield
    (B) public parklands
    (C) previously undeveloped site
    (D) urban area

83. LEED submittal templates require which of the following items? (Choose two.)

    (A) declarant's name
    (B) product manufacturer
    (C) project area
    (D) project location

84. The pre-design phase of a LEED project should include which of the following steps? (Choose three.)

    (A) commissioning mechanical systems
    (B) establishing a project budget
    (C) establishing project goals
    (D) site selection
    (E) testing and balancing mechanical systems
    (F) training maintenance staff

85. Which is true about a CIR submittal?

    (A) CIRs must be submitted as text-based inquiries.
    (B) Drawings and specification sheets must be submitted as attachments.
    (C) It must include a complete project narrative.
    (D) Text is limited to 1000 words.

86. Green building design and construction decisions should be guided by which of the following items? (Choose three.)

    (A) bid cost
    (B) construction documents
    (C) design cost
    (D) energy efficiency
    (E) environmental impact
    (F) indoor environment

87. GBCI is a nonprofit organization that provides which of the following services? (Choose two.)

    (A) accreditation of industry professionals
    (B) certification of sustainable products
    (C) certification of sustainable buildings
    (D) educational programs on sustainability topics

88. Credits can be earned after which of the following project phases? (Choose two.)

    (A) appeal
    (B) design
    (C) certification
    (D) construction
    (E) post-construction

89. What are the short-term benefits of commissioning? (Choose two.)

    (A) assures credit achievement
    (B) decreases initial project cost
    (C) promotes code compliance
    (D) promotes design efficiency
    (E) reduces design and construction time

90. Conventional fossil-based electricity generation results in which of the following emissions? (Choose three.)

    (A) anthropogenic nitrogen oxide
    (B) carbon dioxide
    (C) carbon monoxide
    (D) sulfur dioxide
    (E) VOCs

91. Which strategy helps minimize a site's heat island effect?

    (A) having a high glazing factor
    (B) installing hardscapes with low SRI values
    (C) maximizing the area of site hardscapes
    (D) shading hardscapes with vegetation

92. The BOD, which includes design information necessary to accomplish the owner's project requirements, must contain which of the following? (Choose three.)

    (A) building materials selection
    (B) indoor environmental quality criteria
    (C) mechanical systems descriptions
    (D) process equipment energy consumption information
    (E) references to applicable codes

93. An integrated project team should include which of the following professionals? (Choose two.)

    (A) code official
    (B) energy sustainability consultant
    (C) landscape architect
    (D) product manufacturer
    (E) utility manager

94. A project is considered a major renovation when at least _____ of the building envelope, interior, or mechanical systems is modified.

    (A) 50%
    (B) 60%
    (C) 70%
    (D) 80%

95. A building that will be partially occupied by the owner may pursue LEED CS certification if the building owner occupies no more than _____ of the building's leasable space.

    (A) 25%
    (B) 50%
    (C) 75%
    (D) 80%

96. The LEED EBO&M rating system includes a Best Management Practices prerequisite. This prerequisite would most likely fall under which of the following credit groupings? (Choose two.)

    (A) Energy and Atmosphere
    (B) Indoor Environmental Quality
    (C) Innovation in Design
    (D) Materials and Resources
    (E) Occupant Health and Productivity
    (F) Operational Effectiveness
    (G) Site Management

97. The Project Details section of the LEED project registration form requires which of the following? (Choose three.)

    (A) company names of all team members
    (B) gross area of the building
    (C) list of likely innovation credits to be pursued
    (D) project budget
    (E) site conditions

98. To be eligible for LEED recertification, a project must be

    (A) precertified under the LEED for Homes rating system
    (B) previously certified under the LEED EBO&M rating system
    (C) previously certified as LEED Platinum under the LEED CS rating system
    (D) previously certified at any level other than LEED Platinum under the LEED NC rating system

99. Which of the following can help reduce a building's energy load? (Choose two.)

    (A) reducing the building's heat island effect
    (B) increasing the ventilation rate
    (C) installing heat recovery systems
    (D) flushing out prior to occupancy
    (E) zoning mechanical systems

100. Minimizing which of the following will improve a building's indoor environmental quality?

    (A) acoustical control
    (B) natural lighting
    (C) ventilation rates
    (D) VOC content in building materials

Practice Exam Part Two

1. Ⓐ Ⓑ Ⓒ Ⓓ Ⓔ
2. Ⓐ Ⓑ Ⓒ Ⓓ
3. Ⓐ Ⓑ Ⓒ Ⓓ
4. Ⓐ Ⓑ Ⓒ Ⓓ Ⓔ
5. Ⓐ Ⓑ Ⓒ Ⓓ
6. Ⓐ Ⓑ Ⓒ Ⓓ
7. Ⓐ Ⓑ Ⓒ Ⓓ Ⓔ
8. Ⓐ Ⓑ Ⓒ Ⓓ Ⓔ
9. Ⓐ Ⓑ Ⓒ Ⓓ
10. Ⓐ Ⓑ Ⓒ Ⓓ
11. Ⓐ Ⓑ Ⓒ Ⓓ
12. Ⓐ Ⓑ Ⓒ Ⓓ Ⓔ
13. Ⓐ Ⓑ Ⓒ Ⓓ Ⓔ
14. Ⓐ Ⓑ Ⓒ Ⓓ Ⓔ
15. Ⓐ Ⓑ Ⓒ Ⓓ Ⓔ
16. Ⓐ Ⓑ Ⓒ Ⓓ
17. Ⓐ Ⓑ Ⓒ Ⓓ Ⓔ
18. Ⓐ Ⓑ Ⓒ Ⓓ Ⓔ Ⓕ
19. Ⓐ Ⓑ Ⓒ Ⓓ Ⓔ
20. Ⓐ Ⓑ Ⓒ Ⓓ
21. Ⓐ Ⓑ Ⓒ Ⓓ Ⓔ
22. Ⓐ Ⓑ Ⓒ Ⓓ Ⓔ
23. Ⓐ Ⓑ Ⓒ Ⓓ
24. Ⓐ Ⓑ Ⓒ Ⓓ
25. Ⓐ Ⓑ Ⓒ Ⓓ Ⓔ

26. Ⓐ Ⓑ Ⓒ Ⓓ
27. Ⓐ Ⓑ Ⓒ Ⓓ Ⓔ
28. Ⓐ Ⓑ Ⓒ Ⓓ
29. Ⓐ Ⓑ Ⓒ Ⓓ
30. Ⓐ Ⓑ Ⓒ Ⓓ
31. Ⓐ Ⓑ Ⓒ Ⓓ Ⓔ
32. Ⓐ Ⓑ Ⓒ Ⓓ
33. Ⓐ Ⓑ Ⓒ Ⓓ
34. Ⓐ Ⓑ Ⓒ Ⓓ
35. Ⓐ Ⓑ Ⓒ Ⓓ
36. Ⓐ Ⓑ Ⓒ Ⓓ
37. Ⓐ Ⓑ Ⓒ Ⓓ Ⓔ Ⓕ
38. Ⓐ Ⓑ Ⓒ Ⓓ
39. Ⓐ Ⓑ Ⓒ Ⓓ
40. Ⓐ Ⓑ Ⓒ Ⓓ
41. Ⓐ Ⓑ Ⓒ Ⓓ Ⓔ
42. Ⓐ Ⓑ Ⓒ Ⓓ
43. Ⓐ Ⓑ Ⓒ Ⓓ Ⓔ
44. Ⓐ Ⓑ Ⓒ Ⓓ
45. Ⓐ Ⓑ Ⓒ Ⓓ
46. Ⓐ Ⓑ Ⓒ Ⓓ Ⓔ
47. Ⓐ Ⓑ Ⓒ Ⓓ Ⓔ Ⓕ
48. Ⓐ Ⓑ Ⓒ Ⓓ
49. Ⓐ Ⓑ Ⓒ Ⓓ
50. Ⓐ Ⓑ Ⓒ Ⓓ

51. Ⓐ Ⓑ Ⓒ Ⓓ Ⓔ
52. Ⓐ Ⓑ Ⓒ Ⓓ Ⓔ
53. Ⓐ Ⓑ Ⓒ Ⓓ
54. Ⓐ Ⓑ Ⓒ Ⓓ
55. Ⓐ Ⓑ Ⓒ Ⓓ Ⓔ
56. Ⓐ Ⓑ Ⓒ Ⓓ Ⓔ
57. Ⓐ Ⓑ Ⓒ Ⓓ Ⓔ Ⓕ
58. Ⓐ Ⓑ Ⓒ Ⓓ
59. Ⓐ Ⓑ Ⓒ Ⓓ Ⓔ
60. Ⓐ Ⓑ Ⓒ Ⓓ
61. Ⓐ Ⓑ Ⓒ Ⓓ
62. Ⓐ Ⓑ Ⓒ Ⓓ Ⓔ
63. Ⓐ Ⓑ Ⓒ Ⓓ Ⓔ
64. Ⓐ Ⓑ Ⓒ Ⓓ Ⓔ
65. Ⓐ Ⓑ Ⓒ Ⓓ
66. Ⓐ Ⓑ Ⓒ Ⓓ
67. Ⓐ Ⓑ Ⓒ Ⓓ Ⓔ
68. Ⓐ Ⓑ Ⓒ Ⓓ
69. Ⓐ Ⓑ Ⓒ Ⓓ Ⓔ
70. Ⓐ Ⓑ Ⓒ Ⓓ Ⓔ
71. Ⓐ Ⓑ Ⓒ Ⓓ
72. Ⓐ Ⓑ Ⓒ Ⓓ
73. Ⓐ Ⓑ Ⓒ Ⓓ
74. Ⓐ Ⓑ Ⓒ Ⓓ
75. Ⓐ Ⓑ Ⓒ Ⓓ

LEED ID&C Practice Exam

76. Ⓐ Ⓑ Ⓒ Ⓓ Ⓔ
77. Ⓐ Ⓑ Ⓒ Ⓓ Ⓔ Ⓕ
78. Ⓐ Ⓑ Ⓒ Ⓓ
79. Ⓐ Ⓑ Ⓒ Ⓓ
80. Ⓐ Ⓑ Ⓒ Ⓓ Ⓔ
81. Ⓐ Ⓑ Ⓒ Ⓓ
82. Ⓐ Ⓑ Ⓒ Ⓓ
83. Ⓐ Ⓑ Ⓒ Ⓓ
84. Ⓐ Ⓑ Ⓒ Ⓓ
85. Ⓐ Ⓑ Ⓒ Ⓓ
86. Ⓐ Ⓑ Ⓒ Ⓓ
87. Ⓐ Ⓑ Ⓒ Ⓓ
88. Ⓐ Ⓑ Ⓒ Ⓓ Ⓔ
89. Ⓐ Ⓑ Ⓒ Ⓓ Ⓔ
90. Ⓐ Ⓑ Ⓒ Ⓓ
91. Ⓐ Ⓑ Ⓒ Ⓓ
92. Ⓐ Ⓑ Ⓒ Ⓓ
93. Ⓐ Ⓑ Ⓒ Ⓓ
94. Ⓐ Ⓑ Ⓒ Ⓓ Ⓔ
95. Ⓐ Ⓑ Ⓒ Ⓓ Ⓔ
96. Ⓐ Ⓑ Ⓒ Ⓓ
97. Ⓐ Ⓑ Ⓒ Ⓓ
98. Ⓐ Ⓑ Ⓒ Ⓓ
99. Ⓐ Ⓑ Ⓒ Ⓓ Ⓔ
100. Ⓐ Ⓑ Ⓒ Ⓓ Ⓔ

Practice Exam Part Two

1. Which of the following steps are included in the Credit Interpretation Request process? (Choose three.)

    (A) After the CIR is submitted, the appropriate TAG provides the ruling.
    (B) After USGBC's twice-monthly collection, USGBC member companies provide a consensus-based ruling.
    (C) The LEED project administrator sends the CIR to the local USGBC chapter.
    (D) CIR fees are processed before the CIR is reviewed.
    (E) The ruling is posted to LEED Online no more than 12 business days after it is received.

2. A hotel with tenancy in a casino is seeking LEED certification under the CI rating system and is committed to minimizing the amount of waste sent to landfills. The executive team is currently researching recycling programs to generate revenue and promote their hotels as environmentally responsible. What is the first step in a providing a tenant space with the best plan to reduce and manage waste?

    (A) create a source reduction policy
    (B) designate an area for recycled material separation, collection, and storage
    (C) perform a waste stream audit of the entire building
    (D) perform a waste stream audit of the hotel

3. An entrepreneur plans to lease space within an existing building where she will have the room to add employees and expand her company. Which of the following LEED CI credits would be easier to achieve in a downtown location than in a suburban office park? (Choose two.)

    (A) SS Credit 2, Development Density and Community Connectivity
    (B) SS Credit 3.1, Alternative Transportation—Public Transportation Access
    (C) MR Prerequisite 1, Storage and Collection of Recyclables
    (D) MR Credit 1, Construction, Demolition, and Renovation
    (E) EA Credit 4, Green Power

4. A tenant pursuing LEED certification under the CI rating system in a building that is not LEED certified can achieve SS Credit 1, Site Selection, using which of the following strategies? (Choose three.)

    (A) design a lighting system that reduces energy and eliminates light pollution
    (B) eliminate CFC-based refrigerants in all mechanical systems
    (C) install a stormwater management system that captures rainwater to flush water closets and urinals
    (D) install photovoltaic panels to provide some of the tenant's electricity
    (E) purchase a percentage of the electricity consumed from a Green-e provider

5. Which of the following standards addresses thermal comfort conditions for human occupancy?

    (A) ANSI/ASHRAE 52.2-1999
    (B) ASHRAE 55-2004
    (C) ANSI/ASHRAE 62.1-2007
    (D) ANSI/ASHRAE/IESNA 90.1-2007

6. A design firm is successful at selling sustainable design strategies because its quotes include both the owner's cost to implement the proposed strategies and the owner's projected payback. Which of the following LEED credits have a quantifiable payback? (Choose two.)

    (A) EQ Credit 2, Increased Ventilation
    (B) EA Credit 4, Green Power
    (C) EA Credit 1.1, Optimize Energy Performance—Lighting Power
    (D) WE Credit 1, Water Use Reduction

7. Which of the following strategies can be used to achieve ID Credit 1? (Choose two.)

    (A) create and implement an educational outreach program describing the environmental and human health benefits of green building design
    (B) exceed the ventilation requirements of ANSI/ASHRAE 62.1-2004 by 30%
    (C) implement a credit-earning strategy from another LEED rating system
    (D) include a LEED Accredited Professional on the project team
    (E) include a USGBC member on the project team

8. A tenant space in an existing building is registered for LEED certification under the CI rating system. Since renovating the tenant space may negatively impact the building's overall indoor air quality, the tenant has hired an IAQ manager to oversee construction operations. The IAQ manager has advised the LEED design team to pursue EQ Credit 3.1, Construction IAQ Management Plan—During Construction. Which of the following strategies may be implemented during the construction process to help achieve this credit? (Choose two.)

    (A) do not use air handlers during construction
    (B) install MERV 8 filters in each air handler used during construction
    (C) install MERV 13 filters in each air handler used during construction
    (D) prevent moisture damage to absorptive materials stored on site or installed
    (E) replace all return grille filters before occupancy

9. What information must be gathered to calculate the baseline water usage of plumbing fixtures? (Choose two.)

    (A) full-time equivalent number of occupants
    (B) gallons per flush for installed sewage conveyance fixtures
    (C) usage groups
    (D) flow rates for proposed faucets

10. Every game day, a restaurant within a basketball arena fills to capacity and has continuous bathroom use. The owner, in an effort to promote his business and reduce his water bill, registers for LEED certification under the CI rating system and establishes a baseline water consumption level that meets the requirements of the Energy Policy Act of 1992. To achieve all WE Credits, the aggregate water consumption of the owner's new urinals and water closets must be at least _____ less than the baseline level.

    (A) 30%
    (B) 40%
    (C) 50%
    (D) 60%

11. A building includes four tenant spaces. The building owner intends to replace all of the site's hardscapes with native vegetation and install a new on-site wind turbine to produce a portion of the building's electricity. Under which LEED rating system should the building apply for certification?

    (A) LEED CS
    (B) LEED NC
    (C) LEED EBO&M
    (D) LEED CI

12. Which of the following will satisfy the requirements of EA Credit 4, Green Power? (Choose two.)

    (A) installing daylighting features to reduce the tenant's energy usage by 10%
    (B) installing Energy Star-rated equipment for 70% of eligible equipment
    (C) installing wind turbines that provide 25% of the tenant's electrical consumption for at least four years
    (D) providing at least 50% of the tenant's electricity from renewable sources by engaging in at least a two-year off-site renewable energy contract
    (E) purchasing all of the tenant's energy equivalent in Green-e tradable renewable certificates

13. A pharmaceutical company occupies the top floor of a 40-story building. The floor is equipped with an overhead lighting system. The system has poor aesthetics and no occupant controls, so the owner has decided to remove it, install skylights, and provide everyone in the office with task lighting. Which of the following LEED prerequisites or credits would this affect? (Choose three.)

    (A) EA Prerequisite 1, Fundamental Commissioning of Building Energy Systems
    (B) EA Credit 1.1, Optimize Energy Performance—Lighting Power
    (C) EQ Credit 1, Outdoor Air Delivery Monitoring
    (D) EQ Credit 2, Increased Ventilation
    (E) EQ Credit 6.1, Controllability of Systems—Lighting

14. An existing manufacturing and office facility has achieved LEED Gold certification under the LEED EB rating system. If the owner pursues further certification of his office space under the LEED CI rating system, which of the following will be true? (Choose two.)

    (A) Information regarding areas other than the office can be excluded from the submittal.
    (B) The project team must include a LEED AP.
    (C) The registration fee will be reduced.
    (D) There cannot be any gaps between the performance periods.
    (E) This will be an initial LEED CI certification application.

15. A commercial building to be leased to three retail tenants was designed and constructed over the last two years and finally obtained an occupancy permit last month. If the building has not pursued LEED certification under any rating system and one of the tenants wants to pursue LEED certification, which of the following will be true? (Choose two.)

    (A) All prerequisites of the LEED CI rating system must be achieved.
    (B) The building owner must apply for certification under the LEED NC system before the tenants can apply.
    (C) Design and construction review fees will be refunded if LEED Platinum certification is achieved.
    (D) The LEED EBO&M rating system must be used for certification.
    (E) The tenant cannot apply for LEED certification until two years after occupancy.

16. A hair salon in a mall has achieved a minimum daylight illumination level of 25 foot-candles in 90% of all regularly occupied areas. The space also provides views to the outdoors for 75% of the regularly occupied seated spaces. How many points may be achieved using these strategies?

    (A) 0 points
    (B) 1 point
    (C) 2 points
    (D) 4 points

17. Which of the following prerequisites or credits require use of the FTE calculation? (Choose two.)

    (A) SS Credit 3.2, Alternative Transportation—Bicycle Storage and Changing Rooms
    (B) WE Prerequisite 1, Minimum Water Efficiency
    (C) MR Credit 1.1, Tenant Space—Long-Term Commitment
    (D) EQ Credit 8, Daylight and Views
    (E) EQ Credit 2, Increased Ventilation

18. An insurance agency renovation will include the replacement of all carpeting. Which standards should the interior designer reference to achieve EQ Credit 4.3, Low-Emitting Materials, Flooring Systems? (Choose three.)

    (A) ASTM E1903-97
    (B) Carpet and Rug Institute, Green Label Plus testing program
    (C) Green Seal GC-3
    (D) Green Seal GS-11
    (E) Green Seal GS-36
    (F) SCAQMD Rule 1168

19. Achieving which of the following credits may INCREASE the energy usage of a tenant space? (Choose two.)

    (A) EQ Credit 6, Controllability of Systems
    (B) EQ Credit 1, Outside Air Delivery Monitoring
    (C) EQ Credit 2, Increased Ventilation
    (D) EQ Credit 4, Low-Emitting Materials
    (E) EQ Credit 8, Daylight and Views

20. A LEED project design team intends to achieve an ID Credit for commissioning the building envelope, HVAC, lighting, plumbing, and building automation systems. Before applying for the credit, the project team will submit a Credit Interpretation Request to confirm that they will achieve this credit. Which of the following steps will the team need to take in order to do so? (Choose two.)

    (A) review the LEED CI reference guide for direction
    (B) review the list of previous credit interpretation rulings on LEED online
    (C) submit all Credit Interpretation Requests together
    (D) submit the Credit Interpretation Request online with specification sheets and photographs

21. Which of the following items are excluded from EA Credit 1.4, Optimize Energy Performance—Equipment and Appliances? (Choose two.)

    (A) clothes washers
    (B) food service equipment
    (C) HVAC equipment
    (D) lighting equipment
    (E) office equipment

22. A government-leased office facility would like to reduce its burden on the municipal water supply system. Which of the following strategies could the design-build plumbing contractor implement to achieve this objective? (Choose three.)

    (A) apply for an NPDES permit with the local municipality
    (B) install automatic controls on the bathroom faucets
    (C) install flow reduction aerators in the bathroom faucets
    (D) replace conventional urinals with waterless urinals
    (E) treat all wastewater to tertiary standards on site

23. Once a building has achieved LEED certification under the CS rating system, which of the following is true?

    (A) The building must be recertified under the LEED CS rating system once all the tenant build-outs are complete.
    (B) Tenants must become members of USGBC.
    (C) Tenants must install the items that achieved points under the LEED CS rating system.
    (D) Tenants must pursue LEED certification under the CI rating system.

24. Which of the following statements are true? (Choose two.)

    (A) USGBC inspectors determine the building's certification level upon inspection of the site and/or building.
    (B) The LEED CI reference guide addresses every strategy that can be used to achieve LEED credit.
    (C) Every prerequisite and a minimum amount of credits must be achieved for a project to obtain LEED certification.
    (D) The LEED CI rating system is a third-party green building rating system.

25. Which of the following materials must to be recycled to achieve MR Prerequisite 1, Storage and Collection of Recyclables? (Choose three.)

    (A) batteries
    (B) corrugated cardboard
    (C) fluorescent lightbulbs
    (D) glass
    (E) paper

26. A tenant's energy expenses are not included in the base rent, 80% of all eligible equipment and appliances are Energy Star-rated, and a no smoking policy has been established in the tenant space. How many points may be achieved in the LEED CI rating system?

    (A) 1 points
    (B) 3 points
    (C) 5 points
    (D) 7 points

27. A garden center's offices have been experiencing indoor air quality problems and have recently registered for LEED certification under the CI rating system. Which of the following would help the owner provide the office staff with optimal air quality and earn the project a point? (Choose two.)

    (A) EA Credit 1.4, Optimize Energy Performance—Equipment and Appliances
    (B) EQ Prerequisite 1, Minimum IAQ Performance
    (C) EQ Credit 1, Outdoor Air Delivery Monitoring
    (D) EQ Credit 5, Indoor Chemical and Pollutant Source Control
    (E) EQ Credit 8, Daylight and Views

28. The owner of a 68,000 sq ft aquarium has registered the auditorium for LEED certification under the CI rating system. With the goal of obtaining Platinum certification, the owner hired a commissioning authority and assembled a LEED project team. After review of the construction submittal, the project earned only Gold certification. What action can the team take to pursue LEED Platinum certification? (Choose two.)

    (A) file an appeal for any of the credits that were denied
    (B) register the entire building under the LEED EBO&M rating system
    (C) reregister the project under the LEED CI rating system once applicable changes have been made
    (D) submit a Credit Interpretation Request for any credits that were denied

29. Which of the following defines mitigated stormwater under the LEED CI rating system?

    (A) site precipitation during the performance period
    (B) site precipitation in compliance with the Energy Policy Act of 1992
    (C) site precipitation leaving the site by means of drains
    (D) site precipitation that does not become uncontrolled runoff

30. What is the minimum number of points that a project team must earn to achieve Gold certification under the LEED CI rating system?

    (A) 50 points
    (B) 60 points
    (C) 70 points
    (D) 80 points

31. A room has been designed with deck-to-deck partitions and an exhaust system that operates continuously to maintain negative pressure with respect to the adjacent rooms. This room meets the specifications of which of the following rooms described in which of the following credits or prerequisites? (Choose two.)

    (A) computer server closet—EA Prerequisite 2, Minimum Energy Performance
    (B) high-volume copy room—EQ Credit 5, Indoor Chemical and Pollutant Source Control
    (C) mechanical room—EA Prerequisite 3, Fundamental Refrigerant Measurement
    (D) recycling storage room—MR Prerequisite 1, Storage and Collection of Recyclables
    (E) smoking room—EQ Prerequisite 2, Environmental Tobacco Smoke Control

32. Which of the following water-saving strategies may help achieve an ID credit? (Choose two.)

    (A) install a graywater system for landscape irrigation
    (B) reduce potable water usage by 45%
    (C) install a stormwater treatment system that will remove 80% of the TSS and 40% of the TP
    (D) use nonpotable water for cooling tower makeup

33. A mechanical contracting company is entering into a two-year agreement with the utility company to provide 50% of its energy from a renewable source. What is needed to achieve EA Credit 4, Green Power? (Choose two.)

    (A) verification of the energy efficiency of the mechanical contracting company's mechanical systems
    (B) certificate of renewable energy purchasing agreement
    (C) utility company's certification as Green-e renewable power provider
    (D) mechanical contracting company's compliance with ANSI/ASHRAE/IESNA 90.1-2007

34. How many points can be achieved under ID Credit 1, Innovation in Design?

    (A) 1 point
    (B) 4 points
    (C) 5 points
    (D) as many the LEED project team applies for

35. A donation of used computers to a local school could help achieve what credit?

    (A) MR Credit 2, Construction Waste Management
    (B) MR Credit 4, Recycled Content
    (C) MR Credit 5, Regional Material
    (D) RP Credit 1, Regional Priority

36. A building owner has registered his tenant space for certification under the LEED CI version 2.0 rating system. In 2009 the project team is assembled and ready to proceed with the project. Which version of the rating system must be followed to achieve LEED certification?

    (A) the version that is most applicable to the building
    (B) the version the project was registered under
    (C) the most current version
    (D) any version

37. Which of the following are considered rapidly renewable materials under the LEED CI rating system? (Choose three.)

    (A) bamboo flooring
    (B) cement blocks containing fly ash
    (C) cork flooring
    (D) linoleum flooring
    (E) oak paneling
    (F) salvaged metal fire doors

38. A 5000 sq ft pharmacy in a shopping mall has a five-year lease agreement. The lighting designer has reduced the lighting power densities to 35% below ANSI/ASHRAE/IESNA 90.1-2007. How many points could be achieved for these if the pharmacy pursued LEED certification under the CI rating system?

    (A) 2 points
    (B) 3 points
    (C) 5 points
    (D) 6 points

39. Which of the following industry standards is referenced by EQ Credit 5, Indoor Chemical and Pollutant Source Control?

    (A) ANSI/ASHRAE 52.2-1999
    (B) ASHRAE 55-2004
    (C) ANSI/ASHRAE 62.1-2007
    (D) ANSI/ASHRAE/IESNA 90.1-2007

40. A 1.3 million sq ft multiuse facility is cooled by five centrifugal chillers that use CFC-based refrigerants. If these chillers will not be replaced, what must the project team provide to comply with EA Prerequisite 3, Fundamental Refrigerant Management?

(A) verification that the facility is not regulated by the EPA NPDES requirements
(B) documentation of the refrigeration leakage rate as less than 5% annually and less than 30% over the equipment's entire life cycle
(C) calculations performed by a third party demonstrating that the equipment replacement has more than a 10-year payback
(D) documentation that the chillers are not included in the scope of the LEED CI project

41. For which of the following can an appeal be filed? (Choose two.)

(A) LEED project certification
(B) GBCI's design submittal decision
(C) GBCI's construction submittal decision
(D) a TAG's ruling on a Credit Interpretation Request
(E) GBCI's appeal decision

42. Which of the following strategies would contribute to achieving SS Credit 1, Option 2, Path 5, Heat Island Effect—Roof? (Choose two.)

(A) providing shade for 50% of the site's hardscapes
(B) locating the tenant space within a building that has vegetation on 50% of the roof
(C) providing documentation of maintenance of the roof's vegetated surfaces
(D) locating the tenant space within a building that has a low-sloped roof with a minimum SRI of 29

43. A salvaged door from an abandoned warehouse slated for demolition has been refurbished and is now used as a plan table in an architect's office in a nearby suburb. The use of this door could contribute to achievement of which LEED credits? (Choose two.)

(A) MR Credit 1.2, Building Reuse—Maintain Interior Nonstructural Components
(B) MR Credit 2, Construction Waste Management
(C) MR Credit 3, Materials Reuse
(D) MR Credit 4, Recycled Content
(E) MR Credit 5, Regional Materials

44. A building's facility manager intends to implement programs exceeding the requirements of the LEED CI rating system. Which of the following strategies may earn the project team an ID credit?

(A) configuring all monitoring equipment to generate an alarm when the carbon dioxide level varies by 10% or more from the design's intended conditions

(B) documenting that HVAC & R technicians are accredited by the National Center for Construction Education and Research

(C) having maintenance staff attend quarterly seminars on building maintenance

(D) maintaining the BMPs outlined in the U.S. EPA's *Guidance Specifying Management Measures for Sources of Nonpoint Pollution in Coastal Waters*

45. Which of the following strategies will earn a project team a Regional Priority credit?

(A) purchasing at least 10% of construction materials and furniture from companies within 500 miles of the site

(B) purchasing at least 20% of construction materials and furniture from companies within 500 miles of the site

(C) purchasing at least 50% of construction materials and furniture from companies within 500 miles of the site

(D) pursuing a credit deemed to have additional regional environmental importance by a USGBC regional council

46. Which of the following people can submit his or her LEED AP certificate so that the project will achieve ID Credit 2, LEED Accredited Professional? (Choose three.)

(A) architect
(B) civil engineer
(C) commissioning authority
(D) paint supplier
(E) USGBC inspector

47. The owner of a doctor's office is pursuing LEED CI Platinum certification with the intention of achieving every credit that is financially possible. For the hired commissioning authority to achieve EA Credit 2, Enhanced Commissioning, which of the following must be within the scope of his work? (Choose two.)

(A) building envelope commissioning review

(B) building operations review 10 months after final acceptance

(C) commissioning design review of the OPR both before and after design submittal

(D) commissioning results, findings, and recommendations reported to the owner

(E) contractor submittal review for compliance with the OPR and BOD concurrent with A/E reviews

(F) energy-related systems review and development of a recommissioning plan

LEED ID&C Practice Exam

48. A commuter rail station is located one mile from a building's entrance. A pedestrian path is available from the station to the building. How many points could this arrangement earn in the LEED CI rating system?

    (A) 0 points
    (B) 1 point
    (C) 2 points
    (D) 4 points

49. A project team will conduct a full flush-out after construction and before occupancy. How much outside air must be provided during the procedure to achieve Option 1 of EQ Credit 3.2, Construction IAQ Management Plan—Before Occupancy?

    (A) 0.2 cu ft of outside air per minute per sq ft of floor area
    (B) 0.3 cu ft of outside air per minute per sq ft of floor area
    (C) 14,000 cu ft of outside air per sq ft of floor area
    (D) 16,000 cu ft of outside air per sq ft of floor area

50. Where must windows be located to be considered vision glazing under the LEED CI rating system?

    (A) in the ceiling
    (B) 2.5 ft to 7.5 ft above the floor in exterior walls
    (C) 3.0 ft to 8.0 ft above the floor in interior walls
    (D) 2.0 ft to 7.0 ft above the floor

51. An airport has implemented a no smoking policy throughout the terminal, with smoking permitted only in designated smoking rooms. Which of the following must be included in the smoking room design for compliance with EQ Prerequisite 2, Environmental Tobacco Smoke Control? (Choose three.)

    (A) automatically closing doors
    (B) continuous exhaust to the outdoors, away from intakes and building entry pathways
    (C) deck-to-deck partitions that enclose the entire room
    (D) pressure test verification that the smoking room maintains negative air pressure
    (E) pressure test verification that the smoking room maintains positive air pressure

52. A baseball card collector has recently opened a shop and would like to certify it under the LEED CI rating system. Which of the following credits is the collector eligible to achieve for exemplary performance? (Choose two.)

    (A) SS Credit 2, Development Density and Community Connectivity
    (B) WE Credit 1, Water Use Reduction
    (C) EA Prerequisite 3, Fundamental Refrigerant Management
    (D) MR Credit 1.2, Building Reuse—Maintain Interior Nonstructural Components
    (E) ID Credit 2, LEED Accredited Professional

53. A strip mall tenant would like to register for LEED certification. The owner is happy with her existing space and doesn't want to change more than is necessary for LEED certification. Under which of the following rating systems will the project team most likely recommend she register her space?

    (A) LEED CI
    (B) LEED CS
    (C) LEED EBO&M
    (D) LEED NC

54. Verification that a building's mechanical ventilation system complies with ANSI/ASHRAE 62.1-2007 will help achieve which of the following prerequisites or credits?

    (A) EA Credit 1.3, Optimize Energy Performance—HVAC
    (B) EA Credit 1.4, Optimize Energy Performance—Equipment and Appliances
    (C) EQ Prerequisite 1, Minimum IAQ Performance
    (D) EQ Credit 2, Increased Ventilation

55. Which of the following certifications verify compliance with the Forest Stewardship Council's standards? (Choose two.)

    (A) Center for Resource Solutions certification
    (B) chain of custody certification
    (C) forest management certification
    (D) green wood build certification
    (E) LEED Gold certification

56. An interior designer specializing in incorporating salvaged materials into a building's decor has been hired by a coffee shop owner pursuing LEED certification under the CI rating system. The designer intends to achieve MR Credit 3.2, Materials Reuse—Furniture and Furnishings. Which of the following salvaged or refurbished items can contribute toward this credit? (Choose three.)

    (A) artwork
    (B) chairs
    (C) interior plants
    (D) key cabinets
    (E) window treatments

57. A one-hour photo shop franchise is designing a store with carbon dioxide sensors to help maintain acceptable fresh air ventilation rates and a building automation system that provides optimal start-up and night setback ventilation levels. Which of the following prerequisites and credits will these strategies help achieve? (Choose two.)

    (A) SS Credit 1, Site Selection
    (B) EA Prerequisite 1, Fundamental Commissioning
    (C) EA Prerequisite 2, Minimum Energy Performance
    (D) EQ Prerequisite 1, Minimum IAQ Performance
    (E) EQ Credit 6, Controllability of Systems
    (F) EQ Credit 7, Thermal Comfort

58. After which stage does the project team begin to collect information, perform the required calculations, and write the appropriate policies?

    (A) design submittal
    (B) performance period
    (C) project certification
    (D) project registration

59. Locally manufactured, refurbished plumbing fixtures are being used on a project pursuing LEED certification under the CI rating system. For which of the following credits might this strategy earn a point? (Choose two.)

    (A) WE Credit 1, Water Use Reduction
    (B) MR Credit 1.2, Building Reuse—Maintain Interior Nonstructural Components
    (C) MR Credit 3, Materials Reuse
    (D) MR Credit 4, Recycled Content
    (E) MR Credit 5, Regional Materials

60. Which of the following statements is true of mitigated stormwater?

    (A) impervious paving systems do not affect the amount of mitigated stormwater
    (B) impervious paving systems increase the amount of mitigated stormwater
    (C) pervious paving systems increase the amount of mitigated stormwater
    (D) stormwater retention ponds decrease the amount of mitigated stormwater

61. Locating a tenant space in which of the following buildings will earn a project team the maximum number of points available under SS Credit 1, Site Selection? (Choose two.)

    (A) building compliant with all LEED CI credits involving energy performance design, recycled content, certified wood, and green power
    (B) building compliant with all LEED CI credits involving stormwater design, heat island effect, and light pollution reduction
    (C) building that earned 40 points under the LEED EBO&M rating system
    (D) LEED-registered building

62. Which of the following are benefits of LEED certification under the CI rating system? (Choose two.)

    (A) discount on USGBC materials and courses
    (B) exposure through GBCI website
    (C) optimized energy use for entire life cycle of tenant space
    (D) third-party recognition as environmentally responsible
    (E) use of LEED logo on company website

63. Which of the following are required for LEED certification under the CI rating system? (Choose two.)

    (A) create a plan to recycle mercury-containing lightbulbs
    (B) designate a commissioning authority to perform one commissioning design review of the owner's project requirements and basis of design
    (C) install HVAC equipment in compliance with ASHRAE 55-2004
    (D) provide ventilation rates that meet the requirements of ANSI/ASHRAE 62.1-2007
    (E) reduce water use to a level at least 20% less than the requirements of the Energy Policy Act of 1992

64. A downtown Seattle office building employs 1450 people. Many of the employees drive 30 minutes each way to work and park in a lot four blocks from the office. The company's management is considering pursuing LEED certification and implementing a program to reduce employees' travel time and transportation expenses. Which of the following strategies might earn the team an ID point for this program? (Choose two.)

    (A) demonstrate a quantifiable reduction of automobile usage
    (B) give Fridays off to 20% of the employees
    (C) offer a shuttle service that takes employees to the nearest bus station
    (D) provide secure bicycle racks at the building's entrance for 15% or more of tenant occupants
    (E) verify that home offices have been set up

65. Which of the following results could be expected by implementing a LEED credit addressed by the LEED EBO&M rating system when pursuing LEED certification under the CI rating system?

    (A) achievement of one ID point
    (B) discounted registration fee for future LEED EBO&M certification
    (C) LEED certification under both rating systems
    (D) opportunity to waive a prerequisite of the LEED CI rating system

66. A tenant space was certified under the LEED CI version 2.0 rating system and the tenant intends to register for LEED CI recertification in the future. How much time must elapse from the initial certification before the project may file for recertification?

    (A) 3 months
    (B) 12 months
    (C) 18 months
    (D) the project may not file for recertification

67. Which of the following are requirements of EQ Credit 4.3, Low-Emitting Materials—Flooring Systems? (Choose two.)

    (A) exceed the requirements of Green Seal GC-3 by 10%
    (B) meet the requirements of the Carpet and Rug Institute, Green Label Plus testing program
    (C) purchase carpet systems manufactured within 500 miles of the project site
    (D) use adhesives with a VOC limit less than of 50 grams per liter
    (E) use hard surface floors instead of carpeting in at least 50% space

68. Exceeding the required ventilation rates of ANSI/ASHRAE 62.1-2007 by _____ in all of the occupied spaces will earn _____ toward project certification under the LEED CI rating system.

    (A) 15%, 1 point
    (B) 30%, 1 point
    (C) 15%, a prerequisite
    (D) 30%, a prerequisite

69. A new organic grocery store's construction is complete and about to open. The management has decided that it would like to pursue certification under the LEED CI rating system. How can the project team earn a point for MR Credit 2, Construction Waste Management, if no building retrofits or renovations are anticipated? (Choose two.)

    (A) divert 50% of all future construction waste from the landfill
    (B) divert demolition debris prior to occupancy
    (C) make a minimum dollar donation to a local recycling program
    (D) meet the requirements of LEED NC Prerequisite 1, Construction Activity Pollution Prevention
    (E) provide a written statement that no building retrofits or remodels were performed during the certification process

70. Which of the following are benefits of becoming a LEED AP? (Choose three.)

    (A) achieve one point toward LEED certification for LEED-registered projects
    (B) be listed on the GBCI's directory of LEED APs
    (C) demonstrate a thorough understanding of the LEED CI rating system and LEED process
    (D) receive discounts on LEED reference guides and project registration fees
    (E) obtain access to credit interpretation rulings that have been previously posted

71. If a project team implements only one of the following strategies, which one will earn the project team 10 points?

    (A) exceed the minimum HVAC system performance requirements of ANSI/ASHRAE/IESNA 90.1-2007 by 30%
    (B) install Energy Star-rated equipment for 90% (by rated power) of the eligible equipment
    (C) install HVAC systems that comply with the efficiency requirements of Advanced Buildings™ Core Performance™ Guide Secs. 1.4, 2.9, and 3.10
    (D) install separate control zones for every solar exposure, interior space, private office, and specialty occupancy

72. A civil engineer has increased the volume of a parking lot's mitigated stormwater through use of an open grid pavement system that is less than 50% impervious. What percentage of the parking lot area must meet these specifications to fulfill the requirements of SS Credit 1, Option 2, Path 4, Heat Island Effect—Non-Roof?

    (A) 25%
    (B) 50%
    (C) 75%
    (D) 100%

73. An owner plans to survey all occupants monthly to determine whether they are satisfied with the interior environment. The contracted HVAC engineer has agreed to modify the HVAC system if more than 20% of the occupants are dissatisfied after nine months. Which of the following credits may this effort contribute toward?

    (A) EQ Credit 1, Outdoor Air Delivery Monitoring
    (B) EQ Credit 2, Increased Ventilation
    (C) EQ Credit 6.2, Controllability of Systems—Thermal Comfort
    (D) EQ Credit 7.2, Thermal Comfort—Verification

74. Under the LEED CI rating system, which of the following would be affected by the number of transient occupants in a mall's retail stores?

    (A) bicycle storage requirements
    (B) HVAC zoning requirements
    (C) VOC limits
    (D) water use reduction requirements

75. If an existing HVAC system cannot meet the requirements of the standard referenced in EQ Prerequisite 1, Minimum Indoor Air Quality Performance, then the HVAC system must be modified. How much outside air per person must the system provide throughout the entire occupied cycle?

    (A) 5 cu ft per minute
    (B) 10 cu ft per minute
    (C) 15 cu ft per minute
    (D) 20 cu ft per minute

76. Before beginning the process of LEED certification under the CI rating system, the project team must pay and apply for registration. What information is required with this application? (Choose three.)

    (A) drawings and photos of the project
    (B) LEED CI project scorecard
    (C) list of LEED products that will be used on the project
    (D) overall project narrative
    (E) project team leader's LEED AP certificate

77. To comply with EA Prerequisite 1, Fundamental Commissioning of Building Energy Systems, which of the following items must the commissioning authority's building operation plan address? (Choose three.)

    (A) building automation systems
    (B) building envelope systems
    (C) HVAC systems
    (D) irrigation systems
    (E) lighting systems and safety systems
    (F) plumbing systems

78. Which standard is referenced by SS Credit 1, Option 2, Path 1, Brownfield Redevelopment?

    (A) Stormwater Management for Construction Activities
    (B) ASTM E408-71
    (C) Energy Policy Act of 1992
    (D) ASTM E1903-97, Phase II Environmental Site Assessment

79. Which of the following efficiency levels does the Energy Policy Act of 1992 establish in the LEED CI 2009 rating system?

    (A) building energy usage
    (B) mechanical energy usage
    (C) plumbing fixture water usage
    (D) process water usage

80. After LEED registration, project teams gain access to which of the following items? (Choose three.)

   (A) discounted LEED reference guides
   (B) LEED Online submittal templates
   (C) rulings on previously submitted Credit Interpretation Requests
   (D) project access code
   (E) updated list of LEED APs

81. Which LEED category gives incentive for addressing geographically specific environmental issues?

   (A) Innovation in Design
   (B) Indoor Environmental Quality
   (C) Regional Priority
   (D) Sustainable Sites

82. On-site wastewater must be treated to what level to qualify for a credit?

   (A) graywater
   (B) blackwater
   (C) potable
   (D) tertiary

83. Which of the following must have chain of custody numbers before a forest product can contribute toward earning a LEED point? (Choose two.)

   (A) manufacturer
   (B) distributor
   (C) end user
   (D) installing contractor

84. An integrated project team should be assembled as early as possible and include key stakeholders in the project. Which of these attributes should describe the team? (Choose two.)

   (A) hierarchical
   (B) open
   (C) controlled
   (D) collaborative

85. An educational program on the environmental and human health benefits of green building practices can earn a point within the Innovation in Design category. To achieve this credit, the program must _____. (Choose two.)

   (A) be actively instructional
   (B) be provided by LEED faculty
   (C) include two initiatives
   (D) be approved as an EPP course

86. The use of water-efficient dishwashers and laundry machines can be a factor in meeting the requirements of which of the following?

    (A) WE Prerequisite 1, Water Use Reduction
    (B) WE Credit 1, Water Use Reduction
    (C) ID Credit 1, Path 2, Exemplary Performance
    (D) RP Credit 1, Regional Priority

87. Long-term tenant agreements help in conserving resources, reducing waste, and lessening the environmental impact of relocation and construction activities. How long must a tenant agreement be to count toward achieving a credit?

    (A) 3 years
    (B) 5 years
    (C) 10 years
    (D) 15 years

88. Renewable energy certificates (RECs) represent the positive environmental attributes of power generated from renewable sources. Price premiums may be lower for RECs than for Green-e power purchased from a local utility company, because _____. (Choose three.)

    (A) users are not limited to local sources of RECs
    (B) Green-e power must meet stricter requirements to be considered renewable
    (C) RECs do not need to be delivered to the user
    (D) when demand for Green-e power increases, rates increase
    (E) a supplier of RECs doesn't need to meet the user's real-time electricity needs

89. Which of the following are considered to be regularly occupied spaces? (Choose three.)

    (A) bathroom
    (B) conference room
    (C) classroom
    (D) reception area
    (E) corridor

90. Which of these refrigerants contributes the least to ozone depletion?

    (A) CFC-500
    (B) HCFC-123
    (C) HFC-134a
    (D) CFC-114

91. Development need not increase a site's stormwater discharge rate if strategic site planning occurs. Which of the following strategies can help achieve this goal? (Choose two.)

    (A) on-site retention ponds
    (B) infiltration basins
    (C) high-albedo impervious surfaces
    (D) use of building graywater for landscape irrigation

92. To meet a prerequisite in the LEED CI rating system, lighting power densities must exceed ASHRAE 90.1-2007 requirements by how much?

    (A) 5%
    (B) 10%
    (C) 15%
    (D) 20%

93. A credit can be achieved by including a LEED AP on a project team. In which review phase can material for this credit be submitted? (choose two)

    (A) appeal
    (B) design
    (C) application
    (D) construction

94. How can a project team increase the energy performance of an interior space? (Choose two.)

    (A) installing a control system that provides an unoccupied mode
    (B) installing faucets controlled by motion sensors
    (C) connecting lights to motion sensors
    (D) programming the night setback temperature to be 40°F
    (E) increasing the ventilation rate above ASHRAE 62.1 levels

95. Increased levels of carbon dioxide are typical of which of these kinds of spaces?

    (A) garages
    (B) non-densely occupied spaces
    (C) densely occupied spaces
    (D) smoking rooms
    (E) freshly painted rooms

96. Which of the following is a naturally occurring volatile organic compound (VOC) found in composite wood and agrifiber products?

    (A) urea-formaldehyde
    (B) polychlorinated biphenyl
    (C) chlorofluorocarbons
    (D) fly ash

97. Which of the following are benefits of drip irrigation? (Choose two.)
    (A) quick payback for the cost of system installation
    (B) reduced evapotranspiration rate
    (C) isolation of graywater from human contact
    (D) reduced use of potable water by plumbing fixtures

98. The Energy Star Portfolio Manager allows a project team to _____.
    (A) submit its project as an Energy Star building after LEED certification is achieved
    (B) access a list of energy saving strategies
    (C) view the energy consumption data of local buildings
    (D) benchmark the weather-normalized energy usage of the project against similar occupancies

99. Which of the following can be used to achieve individual occupant control of thermal comfort? (Choose three.)
    (A) operable windows
    (B) thermal comfort surveys
    (C) occupant-adjustable volume dampers on supply grilles
    (D) occupant-adjustable volume dampers on return grilles
    (E) thermostats

100. Which of the following are minimum program requirements (MPRs) that must be met before a project can achieve LEED certification? (Choose three.)
    (A) The enclosed floor area of a commercial interior project must be at least 250 square feet.
    (B) The project's energy usage data must be made available to GBCI and USGBC for five years after certification.
    (C) The project must be recertified every five years to maintain LEED certification status.
    (D) The gross area of the project building must be no more than 40% of the LEED project site area.
    (E) All applicable environmental laws and regulations must be followed.

Practice Exam Part One Solutions

#	Answer	#	Answer	#	Answer
1.	B, D	26.	B, C	51.	C, E
2.	A, C	27.	A, C	52.	A, B, D
3.	A	28.	B, C, D	53.	C
4.	A, B, E	29.	A	54.	B, D
5.	A, B	30.	D	55.	A, D
6.	B, C, E	31.	B, C, E	56.	B, C, D
7.	A, C	32.	C	57.	E, F, G
8.	A, B, C	33.	D	58.	A, C
9.	C, E	34.	A, C	59.	A
10.	C, D	35.	B	60.	A
11.	B, C	36.	D, E	61.	A, E
12.	C, E	37.	C	62.	C
13.	B	38.	A, C, E	63.	D, E
14.	C, D, E	39.	B	64.	D
15.	A	40.	C, D	65.	B
16.	D, E	41.	B, C	66.	D, E
17.	D, E	42.	B, C, E	67.	A, B
18.	C	43.	B, D	68.	C
19.	B	44.	C, D	69.	B
20.	B, D	45.	B, D	70.	B
21.	B, C, D	46.	C	71.	B, D, E
22.	D, E	47.	B	72.	B
23.	A	48.	B, D	73.	D
24.	D	49.	C	74.	B
25.	A, B	50.	B	75.	A

LEED ID&C Practice Exam

76. **A** B C **D** **E**
77. A B C **D** **E**
78. **A** B C D
79. A B C **D**
80. A B **C** D
81. A **B** C D **E**
82. A B C **D**
83. **A** B C D
84. A **B** **C** **D** E F
85. **A** B C D
86. A B C **D** **E** **F**
87. **A** B **C** D
88. **A** B C **D** E
89. A B C **D** **E**
90. **A** **B** C **D** E
91. A B C **D**
92. A **B** **C** D **E**
93. A **B** **C** D E
94. **A** B C D
95. A **B** C D
96. **A** B C D E **F** G
97. A **B** C **D** **E**
98. A **B** C D
99. A B **C** D **E**
100. A B C **D**

Practice Exam Part One Solutions

1. *The answers are:* **(B)** registered employees of USGBC member companies
   (D) registered project team members

   Only members of a LEED project team and employees of a USGBC member company may view posted CIRs.

2. *The answers are:* **(A)** exemplary
   (C) innovative

   To earn points in the Innovation in Design category, project teams can exceed the requirements of an existing LEED credit and earn exemplary performance points, or they can implement innovative performance strategies not addressed by the LEED rating systems.

3. *The answer is:* **(A)** drinking water

   Potable water is water that meets or exceeds EPA drinking water standards and is supplied from wells or municipal water systems.

4. *The answers are:* **(A)** concept
   (B) design development
   (E) construction

   LEED projects benefit from the inclusion of project team members in ongoing commissioning and inspector site visits; however, team member participation is most valuable during the design and construction phases of the project.

5. *The answers are:* **(A)** Administration
   (B) Materials In

   The functional characteristic groups established under the EBO&M rating system are Materials In (includes credits addressing the sustainable purchasing policy of a building), Materials Out, Administration (includes credits addressing the planning and logistics support of operating a high-performance building), Green Cleaning, Site Management, Occupational Health and Productivity, Energy Metrics, and Operational Effectiveness. Sustainable Sites and Water Efficiency are credit categories, not functional characteristic groups. Waste Management is neither a credit category nor a functional characteristic group.

6. *The answers are:* **(B)** adheres to the plan throughout project
   (C) aligns goals with budget
   (E) establishes project goals and expectations

   A project is more likely to stay within budget when the goals and budget are coordinated and when the project team adheres to the plan and frequently checks expenses against the budget. USGBC does not provide guidance for project budgeting. Submitting documentation for fewer credits does not necessarily lead to a successful LEED project budget.

7. *The answers are:* **(A)** accessible roof decks
   (C) non-vehicular, pedestrian-orientated hardscapes

Projects located in urban areas (those buildings with little or no setback) can utilize pedestrian hardscapes, pocket parks, accessible roof decks, plazas, and courtyards to meet the open space requirements.

On-site photovoltaics contribute to on-site renewable energy generation. Pervious parking lots contribute to on-site stormwater mitigation. Landscaping with indigenous plants may reduce the amount of water needed for irrigation. None of these strategies contributes to open space requirements.

8. *The answers are:* **(A)** building orientation
   (B) envelope thermal efficiency
   (C) HVAC system sizing

A building's overall energy consumption can vary depending on building orientation, building location, envelope insulation, fenestration U-values, the size of the HVAC system, and electricity requirements of lighting and appliances. Addressing these factors during the design can result in reduced energy bills for the life of the building. Refrigerant selection affects the building's environmental impact (but not the HVAC system efficiency), and volatile organic compound (VOC) content affects indoor air quality.

9. *The answers are:* **(C)** maintenance
   (E) utilities

Life-cycle costing is an accounting methodology used to evaluate the economic performance of a product or system over its useful life. Life-cycle cost calculations include maintenance and operating costs (including the cost of utilities). Neither the occupants' individual expenses nor the initial cost of an investment factor into a building's life-cycle costs.

10. *The answers are:* **(C)** reduced heat island effect
    (D) reduced light pollution

The primary purpose of increasing a building's daylighting is to reduce the need for electric light. Reducing electric light use results in reduced electricity use in general, thereby reducing energy costs and the carbon dioxide emissions (or air pollution) created by the building. Additionally, statistics show that productivity is markedly improved in day lit buildings compared to buildings that rely heavily on electric lighting.

Light pollution is related to the amount of light transmitted from a building after hours. Daylighting is related to daytime lighting of the interior of the building. A building's heat islands are not affected by daylighting.

11. *The answers are:* **(B)** renovation of part of an owner-occupied building
    (C) tenant infill of an existing building

The LEED for Commercial Interiors (CI) rating system provides the opportunity for tenant spaces and parts of buildings to achieve LEED certification. Major building envelope renovations would be certified under the LEED for New Construction (NC) rating system. Upgrades to the operations and maintenance of an existing facility would be certified under the LEED for Existing Buildings: Operations & Maintenance (EBO&M) rating system.

12. *The answers are:* (C) responsible party
(D) time period

Each policy should identify the individual or team responsible its implementation. Additionally, the time period over which the policy is applicable should be indentified. The time period is not necessarily the same amount of time as the performance period. Note that performance periods only apply to LEED EBO&M projects.

13. *The answer is:* (B) the LEED rating systems

Chain of custody is a procedure to document the status of a product from the point of harvest or extraction to the ultimate consumer end use and can promote sustainable construction. Standard operating procedures (SOPs) are detailed written instructions that document a method with the intention of achieving uniformity. A waste reduction program helps a project team minimize waste by using source reduction, reuse, and recycling. Both SOPs and waste reduction programs promote sustainable operations. Of the answer options, only the LEED rating systems support design, construction, and operations.

14. *The answers are:* (C) primary contact information
(D) project owner information
(E) project type

To register a project for LEED certification, the registrant must provide account login information, primary contact information, project owner information, general project information, payment information, and the project type. Project team member names do not have to be submitted to complete the registration form. LEED projects are not required to have a LEED AP on the team.

15. *The answers are:* (A) achieve certification as a LEED home
(E) market and sell the home

The five basic steps of LEED for Homes are

1. contact a LEED for Homes Certification provider and join the program
2. identify the project team
3. build the home to the stated goals
4. achieve certification as a LEED home
5. market and sell the home

Becoming a USGBC member company reduces the registration fee; however, it is not a basic step.

16. *The answers are:* (D) hydrofluorocarbons
(E) water

Using halons, chlorofluorocarbons, and hydrochlorofluorocarbons in fire suppression systems can lead to ozone depletion. Using water and hydroflourocarbons will have a minimal effect on ozone depletion.

17. *The answer is:* **(D)** total area of the site including constructed and non-constructed areas

The total area within the legal property boundaries of the site is considered the property area. The project site area that includes constructed and non-constructed areas is defined as the development footprint.

18. *The answer is:* **(C)** LEED EBO&M

LEED EBO&M is the only rating system that includes performance periods and a recertification requirement. LEED EBO&M projects may recertify as often as every year, and must recertify at least every five years to maintain their LEED certification status.

19. *The answer is:* **(B)** being pursued and some online documentation has been uploaded

A white check mark indicates that the credit is being pursued and has been assigned to a project member. Some documentation has been uploaded; however, additional info needs to be submitted.

20. *The answers are:* **(B)** energy use reduction
(D) sustainable site use
(E) water use reduction

Remodeled projects typically achieve Building Reuse credits within the Materials and Resources credit category; however, they may have difficulty achieving credits in the Sustainable Sites, Water Efficiency, and Energy and Atmosphere categories. This is because when the site is predetermined, design teams do not have the opportunity to select a more favorable site that would help them achieve those credits. Compared to newer buildings, older buildings usually have less energy-efficient insulation systems, and have less efficient plumbing fixtures.

21. *The answers are:* **(B)** interior finishes
(C) lighting
(D) mechanical distribution

The LEED Core & Shell (CS) rating system is designed to help designers, builders, developers, and new building owners increase the sustainability of a new building's core and shell construction. It covers base building elements and complement the LEED for Commercial Interiors (CI) rating system. Interior space layout, interior finishes, lighting, and mechanical distribution may not be directly controlled by the developer, and therefore if a project team wishes to include these elements in their LEED certification, LEED CS may not be appropriate. Site selection and the building's envelope insulation can be directly controlled by the owner when pursuing LEED CS.

22. *The answers are:* **(D)** project administrators to manage LEED projects
(E) project team members to manage LEED prerequisites and credits

LEED Online does not provide advertising of any sort. Project administrators assign access responsibilities for prerequisites and credits to project team members using LEED Online. Local code officials are unable to view online LEED documentation. The Energy Star Target Finder tool will help a project team analyze anticipated building energy performance.

23. *The answers are:* (A) Heat Island Effect
    (E) Stormwater Design

Green roofs help mitigate stormwater and reduce the roof's heat island effect by increasing evapotranspiration (which has a cooling effect), and increasing the roof's albedo. Light pollution reduction is achieved through strategic lighting design. Site selection credit is achieved by not locating the building, or hardscapes, on the list of prohibited sites.

24. *The answer is:* (D) reduced toxic waste

Because mercury waste is toxic, using light bulbs with low mercury content, long life, and high lumen output will result in reduced toxic waste. Mercury content of lights does not affect light pollution, which refers to the impact of artificial light on night sky visibility.

25. *The answers are:* (A) light to night sky
    (B) light trespass
    (D) stormwater mitigation

Site lighting and stormwater mitigation must be addressed when designing a sustainable site. On-site renewable energy can reduce the building's burden on the power grid and is addressed in the Energy and Atmosphere category. Refrigerants affect ozone depletion and are addressed in the Energy and Atmosphere category.

While important, controlling light trespass and light to night sky are not prerequisites for sustainable site design.

26. *The answers are:* (B) evaluate durability risks of project
    (C) incorporate durability strategies into design

The four basic elements of a durability plan are evaluation of durability risks, incorporation of durability strategies into design, implementation of durability strategies into construction, and completion of a third-party inspection of the implemented durability features.

27. *The answers are:* (A) ensure the technical soundness of the LEED reference guides and training
    (C) resolve issues to maintain consistency across different LEED rating systems

Technical Advisory Groups (TAGs) respond to Credit Interpretation Requests (CIRs) and assist in the development of LEED credits. The Technical Scientific Advisory Committee (TSAC) ensures LEED and its supporting documentation is technically sound while assisting USGBC with complex technical issues.

28. *The answers are:* (B) preindustrial material
(C) rapidly renewable material
(D) regionally extracted material

The LEED EBO&M rating system defines the amount of material that must come from post-consumer, pre-industrial, rapidly renewable, or regionally extracted sources in order to earn credit for ongoing consumable purchases. The Rainforest Alliance certifies food, and is not related to ongoing consumables. Salvaged material use and purchase can contribute to earning credit for the sustainable purchases of durable goods and facility alterations, but not for ongoing consumables.

29. *The answer is:* (A) LEED CI

LEED for Commercial Interiors (CI) addresses tenant spaces within a building. Both LEED for New Construction & Major Renovation (NC) and LEED for Existing Buildings: Operations & Maintenance (EBO&M) are rating systems that apply to entire buildings. LEED for Core & Shell (CS) addresses buildings that are built with no, or limited, interior buildouts.

30. *The answer is:* (D) geographic location

USGBC chapters and regional councils identify which credits are eligible for Regional Priority points based on the needs of each environmental zone. LEED Online determines the region of a project based on its geographic location, which it identifies from the project site's zip code.

31. *The answers are:* (B) construction cost
(C) documentation cost
(E) soft cost

Project teams should consider the potential construction, soft, and documentation costs before committing to pursing a particular LEED credit. The application review cost is established regardless of the number or selection of LEED credits pursued, and is based on the building's floor area. Registration cost is the same for every LEED project, varying only depending on if the project is registered by a member or non-member company.

32. *The answer is:* (C) LEED EBO&M

Credit Interpretation Requests (CIRs) and appeals are components of every LEED rating system. LEED EBO&M is the only rating system that requires the implementation of performance periods.

33. *The answer is:* (D) ANSI/ASHRAE/IESNA 90.1-2007

ANSI/ASHRAE/IESNA 90.1-2007 sets minimum requirements for the energy-efficient design of all buildings except low-rise residential buildings. ANSI/ASHRAE 52.2-1999 addresses air cleaner efficiencies; ASHRAE 55-2004 addresses thermal comfort; and ANSI/ASHRAE 62.1-2007 addresses ventilation.

Practice Exam Part One Solutions

34. *The answers are:* **(A)** Controllability of Systems

(C) Indoor Chemical and Pollutant Source Control

Project teams intending to increase a building's ventilation will have to consider the implications on the building's commissioning, measurement and verification, and energy performance, all of which will be directly affected.

Controllability of Systems relates more to the thermal comfort and lighting of a building than the building's ventilation. Increasing ventilation will not help or prevent a project team from achieving Indoor Chemical and Pollutant Source Control.

35. *The answer is:* **(B)** track the movement of wood products from the forest to the building

A chain of custody document verifies compliance with Forest Stewardship Council (FSC) guidelines for wood products, which requires documentation of every movement of wood products from the forest to the building.

36. *The answers are:* **(D)** manufactured within 500 miles of the project site

(E) permanently installed on the project site

For the purposes of the LEED rating system, regional materials are those permanently installed building components that have been extracted, harvested or recovered, and manufactured within 500 miles of the project site.

Regional materials do not need to be recycled or post-consumer materials, nor do they need to be rapidly renewable (agricultural products that take 10 years or less to grow or raise and can be harvested in an ongoing and sustainable fashion). Forest Stewardship Council (FSC) certification applies only to wood and is not a requirement of regional materials.

37. *The answer is:* **(C)** Materials Reuse

A project team's choice of paint will have little or no effect on materials reuse. Choosing white exterior paint can contribute to reducing a building's heat island effect. Choosing paint with low volatile organic compounds (VOCs) will contribute to earning Low-Emitting Materials credit. Choosing bio-based paint can help a project team earn Rapidly Renewable Materials credit.

38. *The answers are:* **(A)** core credits

(C) innovation credits

(E) prerequisites

All LEED rating systems contain prerequisites, core credits, and innovation credits. Sustainable operations and educational programs may help a project team achieve either a core credit or an innovation credit.

39. *The answer is:* **(B)** door

Computers, office desks, and landscaping equipment are examples of durable goods, which are defined by the LEED reference guides as goods with a useful life of two years or more and that are replaced infrequently. Doors are considered part of the base building equipment.

40. *The answers are:* (C) local climate
    (D) operational costs

Installing larger ventilation systems will minimally impact construction costs, but will significantly increase the energy cost throughout the life cycle of the building. Prior to increasing the ventilation rate of a building, the design winter and summer temperatures and humidity should be considered. Regardless of building location, the interior temperature should typically be between 68°F and 74°F and the relative humidity should be between 50% and 55%. Refrigerant management is not affected by ventilation systems.

41. *The answers are:* (B) global warming potential
    (C) ozone depletion potential

Refrigerants are chemical compounds that, when released to the atmosphere, deteriorate the ozone layer and increase greenhouse gas levels. LEED requires project teams to consider the ozone depletion potential and the global warming potential of refrigerants used in a building's HVAC & R system.

42. *The answers are:* (B) email address
    (C) industry sector
    (E) phone number

There is no cost to set up a personal user account on the USGBC website. The individual does not need to be a USGBC member, have prior LEED project experience, have or intend to have LEED credential, or supply his or her company name.

43. *The answers are:* (B) CIRs reviewed by a TAG for their own project
    (D) CIRs posted prior to project application, for the applicable rating system only

Project teams are required to adhere only to CIRs uploaded to the USGBC website prior to project registration. Adherence is required for those submitted after project registration only if they were submitted by the project team itself. CIR requirements must be adhered to regardless of the project's geographic location; however, project teams are generally only required to follow CIRs for their specific rating system.

44. *The answers are:* (C) achieve the next incremental level of an existing credit
    (D) double the requirements of an existing credit

Innovation in Design points for exemplary performance are earned for going above and beyond existing credit requirements. Alternatively, project teams can earn ID points for achieving the next incremental level of an existing credit if it is specified within the corresponding rating system.

45. *The answers are:* (B) collecting rainwater for sewage conveyance
    (D) using cooling condensate for cooling tower make-up

As with every sustainable strategy, consult all applicable codes prior to implementation. Non-potable water used for cooling tower makeup or sewage conveyance can lead to earning LEED

Practice Exam Part One Solutions

credit. On-site septic tanks do not reduce potable water used for sewage conveyance, and therefore do not contribute to the achievement of a LEED point for water conservation. Code requirements restrict project teams from using blackwater for landscape irrigation.

46. The answer is: (C) LEED credit equivalence submittal

Project teams intending to achieve Innovation in Design credit for innovation (not exemplary performance) must follow the LEED credit equivalence process, which requires the following.

- the proposed innovation credit intent
- the proposed credit requirement for compliance
- the proposed submittal to demonstrate compliance
- a summary of potential design approaches that may be used to meet the requirements

The Credit Interpretation Requests (CIRs) process does not involve the proposition of new credits, and therefore does not require an explanation of a proposed credit requirement. The LEED Online submittal templates do not allow for the proposal of new credits.

47. The answer is: (B) Center for Resource Solutions

The Center for Resource Solutions' Green-e energy program is a voluntary certification and verification program for renewable energy products.

Energy Star's Portfolio Manager is a federal program that helps businesses and individuals protect the environment through energy efficiency. The Department of Energy's mission is to advance the energy security of the United States. There is no such thing as the Center for Research and Development of Green Power.

48. The answers are: (B) International Energy Conservation Code (IECC)
(D) International Mechanical Code (IMC)
(E) International Plumbing Code (IPC)

The International Code Council (ICC) is a consolidated organization that comprises what was formerly the Building Officials and Code Administrators International, Inc. (BOCA), the International Conference of Building Officials (ICBO), and the Southern Building Code Congress International, Inc. (SBCCI). The ICC family of codes includes, but is not limited to, the International Building Code (IBC), the International Fire Code (IFC), the International Plumbing Code (IPC), the International Mechanical Code (IMC), and the International Energy Conservation Code (IECC).

49. The answer is: (C) Companies can be USGBC members.

People can be LEED accredited, buildings can be LEED certified, and companies (not individuals) can be USGBC members. It is not a requirement that a LEED project administrator (or anyone else involved with a LEED project) be a LEED AP.

50. *The answer is:* **(B)** 80%

A project team earning more than 40% but less than 50% of core credits within the LEED rating systems will earn a LEED Certified plaque. They will earn a LEED Silver plaque for earning more than 50% but less than 60% of Core Credits; LEED Gold for earning more than 60% but less than 80%; and LEED Platinum for earning more than 80%.

51. *The answers are:* **(C)** a means to submit a project for review
  (E) the project's final scorecard

Project teams submit their projects to the Green Building Certification Institute (GBCI) for review using the LEED submittal templates, which also generate the final scorecard for LEED projects.

Project team members can manipulate the documentation for only those prerequisites or credits assigned to them. The LEED reference guides contain potential strategies for credit and prerequisite approval. Credit Interpretation Requests are reviewed and submitted at LEED Online, but not though submittal templates.

52. *The answers are:* **(A)** bid
  (B) design
  (D) pre-design

According to the *Sustainable Building Technical Manual*, an environmentally responsive design process includes the following key steps: pre-design, design, bid, construction, and occupancy. While the selection of vendors, consultants, and/or contractors is part of the process, it is not a key step. Rebid and post-design should not occur if an appropriate process has been followed.

53. *The answer is:* **(C)** Projects can earn half points under the Site Selection credit.

LEED EB and LEED EBO&M projects must recertify at least every five years to maintain their LEED certification, but the LEED CI rating system does not have a recertification option. Precertification is only available if a project team is utilizing the LEED for Core & Shell rating system. Prohibiting smoking is an option in every LEED rating system for meeting the Environmental Tobacco Smoke prerequisite; because this is a prerequisite, no points are awarded for compliance.

54. *The answers are:* **(B)** clarify
  (D) denied
  (E) earned

An *earned* ruling requires no additional action from the project team. A *clarify* ruling requires the project team member to address the issues of the project reviewer. A *denied* ruling indicates that the project team member either misunderstood the intent and/or failed to meet the prerequisite or credit. Anticipated and deferred are ways to categorize prerequisites and credits reviewed during the design phase submittal.

55. *The answers are:* **(A)** delegate responsibility and oversee all LEED committee activities
(D) establish and enforce LEED direction and policy

The role of the LEED Steering Committee is to establish and enforce LEED direction and policy as well as to delegate responsibility and oversee all LEED committee activities.

Technical Advisory Groups (TAGs) respond to Credit Interpretation Requests and the GBCI develops and administers the accreditation exams. The Technical Scientific Advisory Committee (TSAC) ensures LEED and its supporting documentation is technically sound.

56. *The answers are:* **(B)** assign credits to project team members
(C) build a project team
(E) submit projects for review

Project administrators can do many things through the LEED Online workspace, including assigning credits to project team members, building the project team, and submitting projects for review. Previously submitted Credit Interpretations Requests (CIRs) can be viewed by USGBC company members as well as LEED project team members; however, credit appeals submitted by other project teams are not available for review. Precertification is a unique aspect of the LEED for Core & Shell rating system.

57. *The answers are:* **(E)** LEED for Homes rating system
(F) sample LEED submittal templates
(G) usgbc.org account

A free download of the LEED for Homes rating system is available at www.usgbc.org/homes. Sample LEED submittal templates are available at LEED Online. Individuals may create a free usgbc.org account.

All Credit Interpretation Requests (CIRs) require a fee be submitted to USGBC before a Technical Advisory Committee (TAG) will review it. The *LEED Reference Guide for Building Design and Construction* is available for purchase at **www.ppi2pass.com/LEED**. USGBC and LEED brochures are available for purchase at www.gbci.org/publications. LEED certification requires project registration, whose fees are described at www.usgbc.org/leedregistration.

58. *The answers are:* **(A)** additional response time may be incurred
(C) it is sent to the LEED Steering Committee

CIRs beyond the expertise of the Technical Advisory Group (TAG) are sent to the LEED Steering Committee and/or relevant LEED Committees for a ruling. Additional time is typically incurred in this situation.

59. *The answer is:* **(A)** LEED project tools

LEED project tools are only available to registered project teams. A directory of all LEED registered and certified projects, as well as posted rulings on Credit Interpretation Requests are available to registered USGBC members. LEED Online does not contain a list of LEED project credit appeals. While the project's submittal templates also become available after the project is registered, sample submittal templates are available to the public.

60. *The answers are:* (A) codes and regulations that address asbestos and water discharge
    (B) codes and regulations that address PCBs and water management

Buildings must be in compliance with federal, state, and local environmental laws and regulations, including but not limited to those addressing asbestos, PCBs, water discharge, and water management. LEED certification can be revoked upon knowledge of noncompliance.

Complying with referenced standards that address sustainable forest management practices, fixture performance requirements for water use, and volatile organic compounds (VOCs) may help achieve LEED credits; however, doing so is not a they are not program requirement.

61. *The answers are:* (A) include members from varying industry sectors
    (E) involve the team in different project phases

An integrated project team actively involves participants from varying industry sectors throughout the project, and meets monthly to review the project status.

The role of the commissioning authority is to verify the mechanical systems are operating as the designer intended. Product vendors may provide valuable insight that promoted the success of the project; however, they are not required on the project team. The purpose of including a LEED AP on a project team is to encourage LEED design integration and to streamline the application and certification processes, but including a LEED AP in not needed to create an integrated project team.

62. *The answer is:* (C) LEED EBO&M

Facilities undergoing minor envelope, interior, or mechanical changes must pursue LEED certification under the EBO&M rating system. Facilities undergoing major envelope, interior, or mechanical changes must pursue LEED certification under the NC rating system. LEED CI is used to certify tenant spaces while LEED CS is used to certify buildings prior to tenant infill.

63. *The answers are:* (D) use rainwater for sewage conveyance or landscape irrigation
    (E) remediate contaminated soil

Soil remediation may help a project team achieve a credit under Brownfield Redevelopment, and using collected rainwater may earn a team points under the Water Efficiency category.

Erosion and sedimentation control, installing HCFC-based HVAC & R equipment, and the storage and collection of recyclables are LEED certification prerequisites (not credits).

64. *The answer is:* (D) To achieve LEED certification, every prerequisite and a minimum number of credits must be achieved.

The first step toward LEED certification is project registration. A fee must be paid when submitting Credit Interpretation Requests (CIRs). There is a LEED rating system for every type of building, including non-commercial buildings.

65. *The answer is:* (B) Optimize Energy Performance

Ground source heat pumps are energy efficient mechanical systems that may help project teams earn the energy performance prerequisite and credit. Ground source heat pumps contain a vapor compression refrigeration cycle, which requires electricity to operate, and therefore it is not a renewable source of energy. They do not generate energy (so they don't contribute to On-Site Renewable Energy credits), nor do they reduce the amount of water used by a project.

66. *The answers are:* (D) offer a volume certification path
    (E) provide a streamlined certification process for large-scale projects

The goals of the Portfolio Program are to assist participants in integrating green building design, construction, and operations into their standard business practices using LEED technical standards and guidance; provide a cost-effective, streamlined certification path for multiple buildings that are nearly identical in design; recognize leaders who are creating market transformation through their commitments and achievements in green building; foster a network of investors, developers, owners, and managers committed to systemically greening their building portfolios; and support participating organizations in fulfilling their sustainability commitments by providing solid performance metrics that can be given to stakeholders.

Submittal templates provide a template of key data for the design team members to compile. The mission of LEED in general is to encourage and accelerate the "global adoption of sustainable green building and development practices through the creation and implementation of universally understood and accepted standards, tools, and performance criteria."

67. *The answers are:* (A) before the design phase
    (B) during the construction phase

The LEED project budget should be addressed before the design phase and throughout the construction phase of the project. LEED project team selection is based on of the created budget.

68. *The answer is:* (C) installing native or adapted plants

Turf grass typically requires sizable amounts of water to sustain it, while native or adapted plants can survive on the amount of rainfall a site naturally receives. Projects with no landscaping are not eligible to earn landscaping credits. Invasive plants, such as weeds, are not considered landscaping by the LEED reference guides.

69. *The answer is:* (B) 4 points

Regionalization is an opportunity for project teams to earn additional points in the Innovation in Design category of each rating system. GBCI determines which six credits are priorities in each region of the United States. Project teams in each region can earn up to four additional points for achieving the Regional Priority credits assigned to its region.

70. *The answer is:* (B) Energy Metrics

Energy Metrics credits focus on the building's energy performance and ozone protection.

Energy and Atmosphere is a credit category within the LEED rating systems, and is not a functional characteristic grouping. Materials Out credits are associated with the sustainable solid waste management policy of a building. Site Management credits address sustainable landscape management practices.

71. *The answers are:* (B) interior moisture loads
   (D) pests
   (E) ultraviolet radiation

The intent of durability planning is to appropriately design and construct high performance buildings that will continue to perform well over time. The principle risks are exterior water, interior moisture, air infiltration, interstitial condensation, heat loss, ultraviolet radiation, pests, and natural disasters.

72. *The answer is:* (B) contact a LEED for Homes provider

The first step for participating in the LEED for Homes program is to contact a LEED for Homes provider and then to register the project with GBCI. Becoming a LEED AP will strengthen your LEED knowledge; however, it is not required to pursue LEED certification in any rating system. Credit Interpretation Requests can be submitted only after a project is registered.

73. *The answer is:* (D) refrigerant leakage

HVAC & R equipment with a relatively short equipment life, a relatively high refrigerant charge, and/or refrigerant leakage will contribute to ozone depletion and global warming.

74. *The answer is:* (B) durability plan

As it is explained in the LEED for Homes rating system, a successful durability plan includes a ranking of durability risks and strategies to minimize the risks, which could include sealing ventilation ducts, installing rodent- and corrosion-proof screens, and using air-sealing pump covers.

The PE exemption form gives project teams the opportunity to follow a streamlined path to achieve certain prerequisites and credits. A building's landscape management plan focuses on ecology and wildlife outside of the building.

75. *The answer is:* (A) ASHRAE 55-2004

ASHRAE 55-2004 was created to establish acceptable indoor thermal environmental conditions. ANSI/ASHRAE 52.2-1999 addresses air cleaner efficiencies; ANSI/ASHRAE 62.1-2007 addresses ventilation; ANSI/ASHRAE/IESNA 90.1-2007 addresses building efficiency.

76. *The answers are:* (A) email address
   (D) organization name
   (E) individual's title

The primary contact of a LEED project is not required to be a LEED AP or have previous LEED project experience.

Practice Exam Part One Solutions

77. *The answers are:* (D) project summary
(E) project team members

The project team and project summary must be confirmed prior to the LEED application process. CIRs and precertification are items available for LEED project administrators; however, they are not required items. The project cost does not need to be confirmed prior to registration.

78. *The answer is:* (A) evapotranspiration

Evapotranspiration is the term used to describe water that is lost through plant transpiration and evaporation from soil.

79. *The answer is:* (D) view a protected electronic version of the LEED reference guide

An online reference guide access code is provided with the purchase of a LEED reference guide. The code provides its owner with electronic access to the reference guide. Project team members can also acquire the access code for 30 days of electronic access to the reference guide from a project team administrator after joining a LEED project.

80. *The answer is:* (C) intent

The structure of LEED prerequisites and credits is considered part of the LEED brand and includes intent, requirements, and potential technologies and strategies. This structure is maintained in each version of the LEED rating system. Economic and environmental benefits, greening opportunities, and submittal requirements are included in the rating systems and/or LEED online.

81. *The answers are:* (B) durable goods
(E) waste stream

The environmental impact of refrigerants and energy consumption are addressed in the Energy and Atmosphere category. Materials and Resources credits address the flow of materials to and from a project site as well as the waste stream generated from a project building.

82. *The answer is:* (D) urban area

Building on a previously developed site helps conserve existing greenfields (undeveloped land). Locating a project in an urban location will provide the occupants with increased opportunities to utilize public transportation and reduce the need to drive to local amenities. LEED prohibits projects from building on public parkland, due to the potential economic and environmental implications.

83. *The answer is:* (A) declarant's name

Every LEED submittal template begins with the following statement: "I, *declarant's name*, from *company name* verify that the information provided below is accurate, to the best of my knowledge."

Product manufacturer info may be required for some documentation; however, it is not a requirement of every submittal. The project location is identified in the LEED registration form.

84. *The answers are:* **(B)** establishing a project budget
(C) establishing project goals
(D) site selection

The pre-design phase of every LEED project includes developing a green vision, project goals, priorities, building program, and budget; assembling a green team; developing partner strategies and project schedules; researching and reviewing local codes, laws, standards; and selecting a project site. Commissioning, testing and balancing, and training are steps of the construction phase.

85. *The answer is:* **(A)** CIRs must be submitted as text-based inquiries.

There is no mechanism available to submit attachments during the CIR process. A complete project narrative is not required, and the CIR is limited to 600 words.

86. *The answers are:* **(D)** energy efficiency
(E) environmental impact
(F) indoor environment

Green building design and construction should be guided primarily by energy efficiency, environmental impact, resource conservation and recycling, indoor environmental quality, and community considerations. Design and bid costs should be considered; however, green building focuses primarily on life-cycle costs. Construction documents are created once the guideline issues have been addressed.

87. *The answers are:* **(A)** accreditation of industry professionals
(C) certification of sustainable buildings

GBCI administers the LEED accreditation for industry professionals, and is responsible for the certification of buildings and parts of buildings. It does not certify products. USGBC provides sustainable education programs.

88. *The answers are:* **(A)** appeal
(D) construction

Credits can be earned during the construction phase of a project. Design credits can be submitted in the design review, but they will only be distinguished as "anticipated" or "denied." Project teams can modify a strategy and resubmit a design submittal credit in the construction phase to change the designation from "denied" to "earned". The construction review must include all design and construction credits that are pursued by the project team. After the construction review, credits are designated as "earned" or "denied." This designation can only be changed by submitting an appeal and paying a fee for each credit appealed. After the credit is appealed, it will be designated "earned" or "denied." The credit cannot be appealed a second time.

Practice Exam Part One Solutions

89. *The answers are:* **(D)** promotes design efficiency

(E) reduces design and construction time

Commissioning increases a project's initial cost, but should reduce its life-cycle cost. Commissioning authorities are not responsible for ensuring a project's compliance with a code. Rather, they verify that review design documents to help eliminate changes during construction or contractors from making design decisions. The commissioning agent's check helps prevent consultant redesign, as well as on-the-job engineering from the contractors, and therefore reduces the overall time spent on design and construction.

90. *The answers are:* **(A)** anthropogenic nitrogen oxide

(B) carbon dioxide

(D) sulfur dioxide

Conventional fossil-fuel generated electricity (such as that from coal-fueled power plants and liquid petroleum) results in the release of carbon dioxide into the atmosphere, which contributes to global warming. Coal-fired electric utilities emit both anthropogenic nitrogen oxide (nitrogen oxide that is a result of human activities, and which is a key contributor to smog) and sulfur dioxide (a key contributor to acid rain).

91. *The answer is:* **(D)** shading hardscapes with vegetation

Heat island effect can be minimized by having high (not low) solar reflectance index (SRI) values, minimizing the area of hardscapes, and shading necessary hardscapes with trees and bushes. The glazing factor is related to daylighting, not heat island effect.

92. *The answers are:* **(B)** indoor environmental quality criteria

(C) mechanical system descriptions

(E) references to applicable codes

Mechanical systems (which include HVAC & R, plumbing, and electrical systems) must be addressed in the basis of design (BOD). The BOD must also establish the procedure for the installed mechanical equipment to achieve the required indoor environmental quality criteria. Applicable codes must also be included to help provide guidance to the installing contractors. Process equipment and building materials may be addressed here; however, it is not a requirement.

93. *The answers are:* **(B)** energy sustainability consultant

(C) landscape architect

Utility managers, product manufacturers, and code officials may be a good resource for sustainability advice; however, they are not directly involved with the decisions of the project, and therefore are not considered part of the integrated project team.

94. *The answer is:* **(A)** 50%

Replacement or upgrade to 50% of the building's envelope (walls, floors, and roof) is considered a major renovation. Replacement or upgrade to 50% of the building's interior (non-structural walls, floor coverings, and drop ceilings) is considered major. Replacement or upgrade to 50% of the building's mechanical systems (HVAC & R, lighting, plumbing) is considered major. The percentages are calculated using either the cost of the renovation or the area being renovated.

95. *The answer is:* **(B)** 50%

Owners can occupy up to 50% of the building's leasable space and still be eligible to pursue LEED certification under the Core & Shell rating system. Buildings in which the owner occupies more than 50% of the leasable floor space must pursue LEED certification under the New Construction rating system.

96. *The answers are:* **(A)** Energy and Atmosphere
(F) Operational Effectiveness

Credits and prerequisites can be grouped by the credit categories designated within the LEED rating systems, or by functional characteristics. The Best Management Practices prerequisite falls within in the Energy and Atmosphere LEED credit category, but can also be grouped as an Operational Effectiveness prerequisite by its functional characteristic. Operational Effectiveness credits and prerequisites support best management practices.

97. *The answers are:* **(B)** gross area of the building
(D) project budget
(E) site conditions

The project's primary contact is the only individual required to submit their company's name. The project team creates a list of possible innovation strategies, which is submitted during the project application phase.

98. *The answer is:* **(B)** previously certified under the LEED EBO&M rating system

LEED for Existing Buildings: Operations & Maintenance is the only rating system that has a recertification option available. Precertification does not exist for the LEED for Schools rating system.

99. *The answers are:* **(C)** installing heat recovery systems
(E) zoning mechanical systems

Increasing the ventilation rate and performing a flush out before occupancy improves indoor air quality but increases the amount of energy used. Reducing a building's heat island effect will not affect the building's energy use.

100. *The answer is:* **(D)** VOC content in building materials

A building's indoor environmental quality can be improved by controlling noise pollution, providing as much natural lighting as possible, and providing adequate ventilation (thereby improving the air quality). Volatile organic compounds (VOCs), which damage lung tissue, should be minimized.

Practice Exam Part Two Solutions

1. **A**, **D**, **E**
2. **D**
3. **A**, **B**
4. **A**, **C**, **D**
5. **B**
6. **C**, **D**
7. **A**, **C**
8. **A**, **D**
9. **A**, **C**
10. **B**
11. **C**
12. **D**, **E**
13. **A**, **B**, **E**
14. **A**
15. **A**
16. **C**
17. **A**, **B**
18. **B**, **E**, **F**
19. **D**
20. **A**, **B**
21. **C**, **D**
22. **B**, **C**, **D**
23. **C**
24. **D**
25. **B**, **D**, **E**
26. **C**
27. **C**, **D**
28. **A**, **B**
29. **D**
30. **D**
31. **B**, **E**
32. **D**
33. **D**
34. **D**
35. **D**
36. **B**
37. **A**, **C**, **D**, **F**
38. **C**
39. **A**
40. **D**
41. **B**, **C**
42. **B**, **D**
43. **C**
44. **C**
45. **D**
46. **A**, **B**, **C**
47. **D**, **E**
48. **A**, **C**
49. **C**
50. **B**
51. **B**, **C**, **D**
52. **B**, **D**
53. **A**
54. **B**, **C**
55. **B**, **C**
56. **B**, **D**, **E**
57. **C**, **D**
58. **D**
59. **D**, **E**
60. **C**
61. **B**, **C**
62. **B**, **D**
63. **E**
64. **A**, **E**
65. **A**
66. **D**
67. **B**, **D**
68. **C**
69. **B**, **E**
70. **A**, **B**, **C**
71. **A**
72. **B**
73. **D**
74. **A**
75. **B**

LEED ID&C Practice Exam

76. A B C D E — filled: A, B, D
77. A B C D E F — filled: A, C, E
78. D
79. C
80. B, C, D
81. C
82. D
83. A, B
84. B, D
85. A, C
86. C
87. C
88. A, C, E
89. B, C, D
90. C
91. A, B
92. B
93. B
94. A, C
95. C
96. A
97. B, C
98. D
99. A, C, E
100. A, B, E

Practice Exam Part Two Solutions

1. *The answers are:* **(A)** After the CIR is submitted, the appropriate TAG provides the ruling.

   (D) CIR fees are processed before the CIR is reviewed.

   (E) The ruling is posted to LEED Online no more than 12 business days after it is received.

GBCI collects the Credit Interpretation Requests twice monthly. However, it is the responsibility of the Technical Advisory Groups (TAGs) to respond to these requests. A TAG has been created for each LEED credit category, and each TAG's members are professionals from around the country who discuss together each Credit Interpretation Request and respond with a ruling. USGBC then sends the TAG's ruling to the project team by email and posts it to LEED Online within 12 business days of receipt from the TAG. The fees associated with the Credit Interpretation Request must be processed prior to the TAG's review and response.

All Credit Interpretation Requests are filed online and sent directly to GBCI. Local USGBC chapters are not directly involved in the LEED submittal process.

2. *The answer is:* **(D)** perform a waste stream audit of the hotel

Before any recycling, reuse, or reduction strategies can be implemented, a waste stream audit is needed to help the team make decisions about the building's waste reduction policy. The policy should include a dedicated area or areas for separation, collection, and storage of recycled materials. The waste stream audit must address at least paper, corrugated cardboard, glass, plastics, and metals for compliance with MR Prerequisite 1, Storage and Collection of Recyclables.

3. *The answers are:* **(A)** SS Credit 2, Development Density and Community Connectivity

   (B) SS Credit 3.1, Alternative Transportation—Public Transportation Access

A project's location often has a substantial impact on the types of LEED credits a team might pursue. Buildings located in urban areas are more likely to have access to public transportation, so a project in such a building would be more likely to pursue SS Credit 3.1. Choosing a location in an urban area can also help minimize the disturbance of existing greenfields by channeling development to urban areas, thus contributing to SS Credit 2.

A drawback to choosing an urban location is that space for recycling and construction waste management is often limited. MR Credit 1, Construction, Demolition, and Renovation falls under the LEED EBO&M rating system, not the LEED CI rating system.

EA Credit 4, Green Power, is achieved through purchase of Green-e power from a utility company. Project location is not a factor.

4. *The answers are:* **(A)** design a lighting system that reduces energy and eliminates light pollution

   (C) install a stormwater management system that captures rainwater to flush water closets and urinals

   (D) install photovoltaic panels to provide some of the tenant's electricity

SS Credit 1, Site Selection, is designed to encourage tenants to select buildings that employ a variety of green strategies. A project can earn points for this credit either by locating the tenant space in a LEED-certified building or by achieving green strategies on the site. Examples of site-specific strategies include stormwater management, heat island reduction, and water-efficient landscaping.

5. *The answer is:* **(B)** ASHRAE 55-2004

ASHRAE 55-2004 addresses thermal comfort for human occupancy; ANSI/ASHRAE 52.2-1999 addresses filter efficiencies; ANSI/ASHRAE 62.1-2007 addresses indoor air quality issues and ventilation rates; and ANSI/ASHRAE/IESNA 90.1-2007 addresses building performance.

6. *The answers are:* **(C)** EA Credit 1.1, Optimize Energy Performance—Lighting Power
   (D) WE Credit 1, Water Use Reduction

Electricity and water use reductions, which can be achieved through EA Credit 1.1 and WE Credit 1, are examples of quantifiable paybacks.

While increasing the outside air rate will provide better indoor air quality, and likely result in higher productivity, this is not quantifiable as a payback. It may also increase the energy usage of the building.

Green power may be purchased from the local utility company for a premium. While this is an environmentally responsible choice, there is not a quantifiable payback associated with this strategy.

7. *The answers are:* **(A)** create and implement an educational outreach program describing the environmental and human health benefits of green building design
   (C) implement a credit-earning strategy from another LEED rating system

Project teams can earn ID points by exceeding the percentage requirements by an amount established by the rating system. Exceeding the ventilation requirements of ANSI/ASHRAE 62.1-2007 by 30% just satisfies the requirements of EQ Credit 2, Increased Ventilation. Implementing a strategy from another LEED rating system is a method of point-earning for ID Credit 1. In addition, a project team may earn one ID point by implementing an actively instructional green building educational outreach program.

Including a LEED AP on the project team earns a point for ID Credit 2, and including a USGBC member on the project team does not earn the project team points under any credit.

8. *The answers are:* **(A)** do not use air handlers during construction
   (D) prevent moisture damage to absorptive materials stored on site or installed

It is good construction practice and a requirement of EQ Credit 3.1 to protect absorptive materials from moisture damage. This may be accomplished by installing MERV 8 filters at each return grille to prevent construction debris from entering the return duct. After construction, the return grille filters must be removed and all air handling filters must be replaced.

Installing MERV 8 or MERV 13 filters in each air handler will not contribute to achieving this credit; however, installing MERV 13 filters in each air handling unit is a requirement of EQ Credit 5, Indoor Chemical and Pollutant Source Control.

9. *The answers are:* (A) full-time equivalent number of occupants
   (C) usage groups

The water consumption for baseline plumbing fixtures is predetermined by the LEED ID&C reference guide. Actual water usage for the selected and installed plumbing fixtures is used to calculate the design water consumption. The full-time equivalent number of occupants is the best estimate of how many people are within the building throughout the week. Usage groups refer to washroom facilities that can be used by only some occupants rather than by all occupants.

10. *The answer is:* (B) 40%

The Energy Policy Act of 1992 establishes minimum efficiency levels for plumbing fixture water usage, and the tenant space's baseline calculation must meet these requirements. Using water saving strategies, the project team must then reduce the project's water use and exceed the baseline calculation by at least 20% to qualify for LEED certification. If the baseline is exceeded by 40% or more in the tenant space, 11 points may be awarded for WE Credit 1, Water Use Reduction.

11. *The answer is:* (C) LEED EBO&M

The project may be LEED certified only under the EBO&M rating system. The CI rating system is not appropriate because certification will be pursued for the entire building. The NC rating system is not appropriate because changes to the project will be limited to site work and the installation of new renewable energy sources; without significant modification to the mechanical systems, building envelope, or building interiors, the project cannot be considered new construction or a major renovation. LEED CS addresses buildings that are built with limited or no interior buildouts.

12. *The answers are:* (D) providing at least 50% of the tenant's electricity from renewable sources by engaging in at least a two-year off-site renewable energy contract
   (E) purchasing all of the tenant's annual energy equivalent in Green-e tradable renewable certificates

On-site renewable energy, such as from wind turbines, may earn a point under the ID category, but it does not qualify for a credit under the EA category. Daylighting features that reduce energy consumption help the project achieve EA Prerequisite 2, Minimum Energy Performance, but this strategy does not qualify as green power. Installing Energy Star-rated equipment for 70% of eligible equipment would earn the project team one point for EA Credit 1.4, Optimize Energy Performance—Equipment and Appliances.

13. *The answers are:* (A) EA Prerequisite 1, Fundamental Commissioning of Building Energy Systems
    (B) EA Credit 1, Optimize Energy Performance—Lighting Power
    (E) EQ Credit 6.1, Controllability of Systems—Lighting

EA Prerequisite 1, Fundamental Commissioning, requires the lighting systems to be commissioned. Skylights may increase the building's energy use, but they also provide natural light and may decrease the energy used by the lighting system, which is addressed by EA Credit 1, Optimize Energy Performance. Task lighting for at least 50% of the building's occupants will achieve EQ Credit 6.1, Controllability of Systems: Lights.

Ventilation rates are based on the number of building occupants. Monitoring the outside air and increasing ventilation rates are related to indoor air quality, not the installation of skylights or task lighting.

14. *The answers are:* (A) Information regarding areas other than the office can be excluded from the submittal.
    (E) This will be an initial LEED CI certification application.

The LEED CI rating system does not require recertification of a project to maintain its LEED status, so this will be considered an initial certification and the registration fee will be the same regardless of previous certification.

The LEED CI rating system applies to parts of buildings, so data from the rest of the building will not be necessary.

Performance periods are set up in the LEED EBO&M rating system, but are not necessary for the LEED CI rating system. Furthermore, the project team can benefit from having a LEED AP, but it is not a requirement of the rating system.

15. *The answers are:* (A) All prerequisites of the LEED CI rating system must be achieved.
    (C) Design and construction review fees will be refunded if LEED Platinum certification is achieved.

Tenant spaces must pursue LEED certification under the CI rating system, and thus all CI prerequisites must be achieved. Any project earning LEED Platinum certification will be refunded its design and construction review fees. If the building owner wanted to pursue certification of the entire building, he would have to do so under the EBO&M rating system. Projects pursuing LEED certification under the EBO&M rating system are penalized if they have not achieved LEED certification under the LEED NC rating system first.

16. *The answer is:* (C) 2 points

Providing a minimum illumination level of 25 footcandles in 75% of the regularly occupied spaces achieves one point under EQ Credit 8.1, Daylight and Views—Daylight, and providing this in 90% of the regularly occupied spaces achieves two points. 90% of the regularly occupied seated spaces must have a view to the outdoors for a point to be achieved.

Practice Exam Part Two Solutions

17. *The answers are:* (A) SS Credit 3.2, Alternative Transportation—Bicycle Storage and Changing Rooms
    (B) WE Prerequisite 1, Minimum Water Efficiency

SS Credit 3.2, Alternative Transportation—Bicycle Storage and Changing Rooms, requires the full-time equivalency (FTE) calculation to determine the number of bike racks required. The WE prerequisite and credits require the calculation of the FTE occupants to determine the plumbing fixture count and water consumption.

EQ Credit 8, Daylight and Views, is based on floor area of regularly occupied spaces, not the number of FTE occupants, and MR Credit 1.1, Tenant Space, does not require the FTE calculation.

EQ Credit 2, Increased Ventilation, requires that the mechanical ventilation system provide at least 30% more fresh air than is required by ANSI/ASHRAE 62.1-2007, which outlines a "Ventilation Rate Procedure" with respect to the amount of outdoor air required to be provided to a building. The procedure is based on space type and application, occupancy level, and floor area. This occupant load is based on predetermined values, listed in Table 6-1 of the referenced standard. The occupancy level is not obtained using LEED's FTE calculation.

18. *The answers are:* (B) Carpet and Rug Institute, Green Label Plus testing program
    (E) Green Seal GS-36
    (F) SCAQMD Rule 1168

Any carpet products must meet their respective standards to qualify for EQ Credit 4.3, Low-Emitting Materials, Flooring Systems. The Carpet and Rug Institute identifies low-emitting carpet products so consumers can purchase carpet with minimal negative impact on the indoor air quality. Carpet adhesives must have volatile organic compounds not exceeding the limits set by SCAQMD Rule 1168. Any aerosol adhesives used must meet the requirements of Green Seal GS-36.

ASTM E1903-97 concerns testing and mitigation of contaminated soils. Green Seal GS-11 and GC-3 address volatile organic compound (VOC) limits in paints.

19. *The answers are:* (C) EQ Credit 2, Increased Ventilation
    (E) EQ Credit 8, Daylight and Views

EQ Credit 8 requires providing additional outside air and glass in exterior walls, which can improve the interior comfort of the space for the building's occupants, but it may increase the building's energy usage. Similarly, the requirements of EQ Credit 2 may increase the building's energy usage.

Providing control of the building's zones, as in EQ Credit 6.1, Controllability of Systems—Lighting, and monitoring the outside air provided, as in EQ Credit 1, Outside Air Delivery Monitoring, can help minimize the building's energy usage. Using low-emitting materials, as required in EQ Credit 4, will not generally impact energy use.

20. *The answers are:* (A) review the LEED CI reference guide for direction
    (B) review the list of previous credit interpretation rulings on LEED Online

Before the submitting a Credit Interpretation Request, a project team member should review the LEED CI reference guide to determine if the question has already been addressed, and then LEED Online for rulings on past Credit Interpretation Requests. If the question has not been previously asked, the LEED team member should submit a brief text-based question. The question is limited to 600 words, only one question per request is permitted, and there is no mechanism for submitting specification sheets, drawings, photographs, or attachments of any kind.

21. *The answers are:* (C) HVAC equipment
    (D) lighting equipment

HVAC, lighting, and building envelope equipment are excluded from meeting the requirements of EA Credit 1.4, Optimize Energy Performance—Equipment and Appliances. The credit includes appliances, office equipment, electronics, and food service equipment.

22. *The answers are:* (B) install automatic controls on the bathroom faucets
    (C) install flow reduction aerators in the bathroom faucets
    (D) replace conventional urinals with waterless urinals

Decreasing a building's water consumption will decrease the building's water bill and the demand it places on the community's municipal water system. Reduced potable water demand helps protect aquatic ecosystems and reduce the amount of chemicals required to treat the discharge water. Automatic controls or aerators on faucets, low flow or waterless urinals, and graywater systems are all strategies that can be implemented to reduce the amount of potable water required by the building.

National Pollution Discharge Elimination System (NPDES) permit compliance is typically required in most municipalities. The program controls water pollution by regulating point sources that discharge pollutants into U.S. waters. Treating wastewater to tertiary standards on site does not decrease the amount of water supplied to the building.

23. *The answer is:* (C) Tenants must install the items that achieved points under the LEED CS rating system.

Once a building has achieved LEED certification under the CS rating system, the tenants must install the items that achieved LEED credits or prerequisites. They are not required to pursue LEED for CI certification or become member companies of USGBC. There is no recertification program required for the LEED CS rating system.

24. *The answers are:* (C) Every prerequisite and a minimum amount of credits must be achieved for a project to obtain LEED certification.
    (D) The LEED CI rating system is a third-party green building rating system.

USGBC does not employ building inspectors who verify a building's compliance with the LEED rating systems. Certification is based upon the documentation submitted by the project team.

The LEED reference guides do not provide exhaustive lists of green building strategies for meeting the requirements of each credit. As new technologies are introduced, the ID category may be used to explore and achieve credits for progressive strategies not specifically addressed by the LEED reference guides.

25. *The answers are:* (B) corrugated cardboard
    (D) glass
    (E) paper

Every building registered under the LEED CI rating system must recycle at least paper, glass, corrugated cardboard, plastics, and metals. Furthermore, an easily accessible area must be provided to house the separation, collection, and storage of recyclables. Recycling batteries and fluorescent lightbulbs is not required for this prerequisite, but it is good sustainable design practice, and is required in some municipalities.

26. *The answer is:* (C) 5 points

Project teams may earn three points for EA Credit 3, Measurement and Verification, by requiring the tenant to pay for the tenant's energy usage. Two points may be awarded for EA Credit 1.4, Optimize Energy Performance—Equipment and Appliances, if at least 77% of eligible equipment and appliances are Energy Star-rated. Prohibiting indoor smoking meets the requirements of EQ Prerequisite 2, Environmental Tobacco Smoke (ETS) Control, and so earns no points.

27. *The answers are:* (C) EQ Credit 1, Outdoor Air Delivery Monitoring
    (D) EQ Credit 5, Indoor Chemical and Pollutant Source Control

EQ Credit 5 requires installing MERV 13 filters capable of filtering 100% of both the return air and outside air. This can enhance the indoor air quality of the space. Meeting EQ Credit 1 requires monitoring the outside air and alerting occupants when the system is not performing as designed so that adjustments can then be made.

Providing the office with natural light and views from the outdoors will improve the indoor environmental quality (which refers to temperature, humidity, lighting, and noise), not the indoor air quality. LEED projects are required to achieve every prerequisite to be eligible for LEED certification. Points are not awarded for prerequisites. The intent of EA Credit 1.4 is to reduce the impacts of excessive energy use, so implementing an outdoor air delivery system might negatively affect EA Credit 1.4.

28. *The answers are:* (A) file an appeal for any of the credits that were denied
    (B) register the entire building under the LEED EBO&M rating system

The team can submit an appeal for each credit within 25 business days of the final review. Additionally, the team can register the entire building and attempt to achieve LEED Platinum certification under the LEED EBO&M rating system.

Credit Interpretation Requests must be submitted before the final review. There is no opportunity to reregister projects under the LEED CI rating system.

29. *The answer is:* (D) site precipitation that does not become uncontrolled runoff

Under the LEED rating systems, mitigated stormwater is considered to be site precipitation that does not become uncontrolled runoff. Factors affecting stormwater mitigation include site perviousness, stormwater management practices, and on-site capture and reuse of rainwater.

Runoff is defined as stormwater leaving the site by means of uncontrolled surface streams, rivers, drains, or sewers. Rainwater cannot be considered as potable or in compliance with the Energy Policy Act of 1992. There are no performance periods under the LEED CI rating system.

30. *The answer is:* (C) 60 points

Every prerequisite of the LEED CI rating system must be achieved for a project to be LEED certified. After meeting the prerequisites, there are four levels of LEED certification. The required points for each level are as follows.

Certified	40–49 points
Silver	50–59 points
Gold	60–79 points
Platinum	80–110 points

31. *The answers are:* (B) high-volume copy room—EQ Credit 5, Indoor Chemical and Pollutant Source Control
(E) smoking room—EQ Prerequisite 2, Environmental Tobacco Smoke Control

Smoking rooms and high-volume copy rooms must be designed with deck-to-deck partitions and negative pressure from continuous exhaust. Indoor air quality issues must be addressed in mechanical rooms, computer server closets, and recycling rooms, but exhaust fans and deck-to-deck partitions are not required.

32. *The answers are:* (B) reduce potable water usage by 45%
(D) use nonpotable water for cooling tower makeup

Supplying cooling tower makeup water from a nonpotable source can achieve an ID credit. Nonpotable water is defined by USGBC as water that is not suitable for drinking without treatment. Rainwater and condensate are examples of nonpotable water.

SS Credit 1, Site Selection, addresses stormwater treatment systems and graywater systems for landscape irrigation.

33. *The answers are:* (B) certificate of renewable energy purchasing agreement
(C) utility company's certification as a certified Green-e renewable power provider

To achieve EA Credit 4, Green Power, electricity must be purchased from a certified Green-e renewable power provider and the LEED project team must submit documentation of the renewable energy purchasing agreement.

Energy efficiencies of mechanical systems and compliance with ANSI/ASHRAE/IESNA 90.1-2007 address the energy performance of the tenant space. EA Credit 4, Green Power, does not require that a tenant space be energy efficient.

34. *The answer is:* **(D)** 5 points

The purpose of the Innovation in Design category is to give the LEED project the opportunity to achieve points for green building strategies that are not specifically addressed by the LEED CI rating system. The category lets the LEED project team earn points for up to five strategies not addressed by the rating system. Only one ID point can be earned for each strategy.

35. *The answer is:* **(A)** MR Credit 2, Construction Waste Management

Donated materials are considered to be diverted from the waste stream and thus can help achieve MR Credit 2, Construction Waste Management. The donation need not be to a local entity. Materials that are sold can also count toward this credit.

36. *The answer is:* **(B)** the version the project was registered under

The version the project team registers under must be followed to achieve LEED certification. The project team may, however, upgrade to the current version by reregistering the project. Check with GBCI for associated fees and rules for project version upgrades.

37. *The answers are:* **(A)** bamboo flooring
      (C) cork flooring
      (D) linoleum flooring

The LEED CI rating system defines rapidly renewable materials as products that can be grown and harvested within a 10-year cycle. This definition includes bamboo, cork, and linoleum. Oak trees do not reach maturity within 10 years. Salvaged materials do not qualify for this credit unless they have been created from a renewable source. Fly ash is one of the residues generated in the combustion of coal. It is considered a pre-consumer recycled material, not a rapidly renewable resource.

38. *The answer is:* **(C)** 5 points

Project teams are eligible to earn one point for 10-year lease agreements, and no points for five-year leases. Reducing the lighting power density to 35% below the ANSI/ASHRAE/IESNA 90.1-2007 will earn the team five points. One ID point may be achieved if the lighting power density is 40% below the allowable limits of the standard.

39. *The answer is:* (A) ANSI/ASHRAE 52.2-1999

ANSI/ASHRAE 52.2-1999 establishes a procedure for determining the performance of filters with respect to particle size. EQ Credit 5, Indoor Chemical and Pollutant Source Control, requires that appropriate filtration devices be installed for optimal indoor air quality.

ASHRAE 55-2004 details acceptable indoor thermal comfort levels for a percentage of the building's occupants. ANSI/ASHRAE 62.1-2007 addresses the required ventilation rates for indoor air quality. ANSI/ASHRAE/IESNA 90.1-2007 provides the minimum requirements of an energy-efficient building design.

40. *The answer is:* (D) documentation that the chillers are not included in the scope of the LEED CI project

LEED CI projects are exempt from the requirements of the Montreal Protocol if the HVAC & R equipment is not installed or renovated as part of the project scope.

The Montreal Protocol addresses the ozone-depleting properties of chlorofluorocarbons (CFCs) by requiring the phaseout of all CFC-based refrigerants. The LEED CI rating system requires that equipment using CFC refrigerants be replaced with equipment that does not use CFC refrigerants unless a 10-year payback, calculated by a third party, cannot be achieved. The EPA Clean Air Act regulates the use and recycling of ozone-depleting compounds. Refrigeration leakage must also be addressed.

41. *The answers are:* (B) GBCI's design submittal decision

(C) GBCI's construction submittal decision

The project team has 25 business days to appeal GBCI's design or construction submittal decision. Once a project is certified, appeals cannot be submitted. Appeals and CIR rulings cannot be appealed.

42. *The answers are:* (B) locating the tenant space within a building that has vegetation on 50% of the roof

(D) locating the tenant space within a building that has a low-sloped roof with a minimum SRI of 29

Site hardscapes are addressed by SS Credit 1, Option 2, Path 5, Heat Island Effect—Non-Roof. It is good design practice to maintain the installed vegetated roof, but it will not help the project team achieve this credit.

43. *The answers are:* (C) MR Credit 3, Materials Reuse

(E) MR Credit 5, Regional Materials

Under MR Credits 3 and 5, building construction items that are not fixed, such as doors, are considered materials. Walls, ceilings, and flooring, which are covered in MR Credit 1.2, Building Reuse—Maintain Interior Nonstructural Components, are considered fixed. Under MR Credit 5, items that have been extracted and/or manufactured within 500 miles of the project site are considered regional materials.

MR Credit 2, Construction Waste Management, applies to the construction process of the CI project scope. Salvaged and refurbished materials are excluded from MR Credit 4, Recycled Content.

Practice Exam Part Two Solutions

44. The answer is: (C) having maintenance staff attend quarterly seminars on building maintenance

Having maintenance staff attend quarterly seminars on building maintenance goes beyond the scope of the LEED CI rating system, so this could potentially qualify a project team for an ID point. To confirm eligibility, it would be wise for LEED project to submit a Credit Interpretation Request.

EQ Credit 1, Outdoor Air Delivery Monitoring, requires configuration of all monitoring equipment to generate an alarm when the carbon dioxide level varies by 10% or more from the design's intended conditions. Accreditation by the National Center for Construction Education and Research is a common requirement of HVAC & R technicians, so it is unlikely that it would earn a project team an ID point. Maintaining the best management practices (BMPs) outlined in the U.S. EPA's *Guidance Specifying Management Measures for Sources of Nonpoint Pollution in Coastal Waters* is a requirement of SS Credit 1, Option 2, Path 3, Stormwater Design—Quality Control.

45. The answer is: (D) pursuing a credit deemed to have additional regional environmental importance by a USGBC regional council

Each regional authority identifies six items of regional importance that qualify as Regional Priority credits. The LEED project team can pursue or achieve a LEED point for up to four of these approved items.

Purchasing project items locally does not necessarily achieve Regional Priority credits. These points are available under MR Credit 5, Regional Materials.

46. The answers are: (A) architect
(B) civil engineer
(C) commissioning authority

To achieve the ID Credit 2, LEED Accredited Professional, at least one principal member of the LEED project team must submit his or her LEED Accredited Professional certificate. Project suppliers are not considered principal participants of the design team. USGBC does not employ inspectors for LEED CI projects.

There are currently three exam tracks: LEED NC, LEED CI, and LEED EBO&M. The LEED AP certificate denotes which exam track was taken to achieve accreditation, but the project team can earn a point for ID Credit 2 regardless of the exam track used by the LEED AP for the project.

47. The answers are: (D) commissioning results, findings, and recommendations reported to the owner
(E) contractor submittal review for compliance with the OPR and BOD concurrent with A/E reviews

For EA Credit 2, Enhanced Commissioning, the commissioning authority must

- conduct at least one commissioning design review of the owner's project requirements (OPR), basis of design (BOD), and design documents before the start of the mid-construction documents phase

- back-check the review comments in the subsequent design submission
- review contractor submittals applicable to systems being commissioned for compliance with the OPR and BOD (concurrent with A/E reviews)
- report results, findings, and recommendations directly to the owner

While it is not required for EA Credit 2, an owner can attain significant financial savings and a reduced risk of poor indoor air quality by including building envelope commissioning.

48. The answer is: **(A)** 0 points

The rail station must be within a half mile of the building entrance to qualify for SS Credit 3.1, Alternative Transportation—Public Transportation Access, so no points would be awarded for pedestrian access to this station. The points could be achieved if a shuttle were provided between the building and the rail station.

49. The answer is: **(C)** 14,000 cu ft of outside air per sq ft of floor area

The indoor temperature must be at least 60°F and the indoor humidity must be 60% during the flush-out cycle and all construction must be complete prior to it. It is acceptable for furniture and office equipment to be installed during the flush-out cycle, which does not have to be continuous.

50. The answer is: **(B)** 2.5 ft to 7.5 ft above the floor in exterior walls

The LEED CI reference guide defines vision glazing as glazing that provides views of outdoor landscapes to occupants for vertical windows between 2.5 ft and 7.5 ft above the floor. Windows below 2.5 ft and above 7.5 ft, including daylight glazing, skylights, and roof monitors, do not count as vision glazing under the LEED CI rating system.

51. The answers are: **(B)** continuous exhaust to the outdoors, away from intakes and building entry pathways
  (C) deck-to-deck partitions that enclose the entire room
  (D) pressure test verification that the smoking room maintains negative air pressure

All interior smoking rooms must be designed with a continuously operating exhaust fan that provides an average minimum negative pressure of 5 Pa. A pressure test must verify the differential pressure of the smoking room with respect to the surrounding rooms. The smoking room must also have deck-to-deck partitions to prevent environmental tobacco smoke from migrating to other parts of the building. Designing the room with automatic doors is good design practice; however, it is not required by this prerequisite.

52. The answers are: **(B)** WE Credit 1, Water Use Reduction
  (D) MR Credit 1.2, Building Reuse—Maintain Interior Nonstructural Components

SS Credit 2, Development Density and Community Connectivity, is not eligible for exemplary performance points, nor are any LEED prerequisites. The number of LEED APs included on the LEED project team does not change the team's eligibility for point earning, so ID Credit 2, LEED Accredited Professional, is also not eligible for an exemplary performance point.

53. *The answer is:* **(A)** LEED CI

LEED CI applies to tenant spaces. The LEED NC rating system addresses major renovation projects such as replacement of all the mechanical systems and installation of skylights throughout the entire building. The LEED EBO&M rating system is a whole-building rating system and is not applicable to individual tenant spaces. The LEED CS rating system addresses the design and construction issues that are under the direct control of the owner/developer and is designed to compliment the LEED CI rating system.

It is possible for more than one rating system to apply to a project. For a project to be eligible for a rating system, the design team must verify that each prerequisite can be achieved and that the minimum amount of credits required for LEED certification can be obtained. If more than one rating system applies, the design team can choose which rating system will be followed.

54. *The answer is:* **(C)** EQ Prerequisite 1, Minimum IAQ Performance

ANSI/ASHRAE 62.1-2007 outlines the ventilation requirements of commercial buildings. Project teams must comply with this standard (or local codes if they are more stringent) for LEED certification under the CI rating system.

55. *The answers are:* **(B)** chain of custody certification
(C) forest management certification

FSC forest management certification is awarded to responsible forest managers upon successful completion of a forestry practices audit. After successful completion of an audit confirming proper use of FSC wood, name, and logo, chain of custody certification is awarded to companies that process, manufacture, or sell products that contain FSC-certified wood.

A project earning LEED Gold certification will not have necessarily have complied with the Forest Stewardship Council's standards. The Center for Resource Solutions Green-e products certification defines renewable resources under EA Credit 4, Green Power. Green wood build is not a kind of certification.

56. *The answers are:* **(B)** chairs
(D) key cabinets
(E) window treatments

Only items included in CSI MasterFormat™, Division 12, can count toward MR Credit 3.2, Materials Reuse—Furniture and Furnishings. Artwork, interior plants, and musical instruments cannot be included in the credit's calculation.

57. *The answers are:* **(C)** EA Prerequisite 2, Minimum Energy Performance
(D) EQ Prerequisite 1, Minimum IAQ Performance

Carbon dioxide sensors and a building automation system providing optimal start-up and night setback ventilation levels are energy saving strategies that can contribute toward EA

Prerequisite 2, Minimum Energy Performance. Maintaining fresh air ventilation rate can contribute toward achieving EQ Prerequisite 1, Minimum IAQ Performance.

These efforts meet minimum LEED certification requirements only, and will not contribute toward achieving any LEED credits under the CI rating system.

58. The answer is: (D) project registration

Once a project is registered, the LEED project team begins to collect information as required by all prerequisites and the credits being pursued. Once a project is reviewed, the LEED project team can either accept the decision or appeal the credits that have been denied. Performance periods apply only to projects pursuing LEED EBO&M certification.

59. The answers are: (D) MR Credit 4, Recycled Content
(E) MR Credit 5, Regional Materials

Mechanical, electrical, and plumbing components can be used toward achieving MR Credit 3, Materials Reuse. The LEED CI reference guide defines a regional product as one manufactured within 500 miles of the project. Specifying local products can help a project team achieve MR Credit 5, Regional Materials.

Refurbished plumbing fixtures may contribute toward achieving points in the Water Efficiency category if they meet the water consumption requirements of WE Credit 1, Water Use Reduction, but the credits are not awarded for fixture reuse. MR Credit 1.2, Building Reuse, applies only to interior walls, floors, and ceiling systems, not to plumbing fixtures.

60. The answer is: (C) pervious paving systems increase the amount of mitigated stormwater

As addressed in SS Credit 1, Option 2, Paths 2 and 3, pervious surfaces allow moisture to pass through and soak into the earth below. Impervious surfaces do not allow moisture to pass through, so the moisture becomes runoff. Mitigated water is the volume of water that falls on the site and does not become runoff. Stormwater retention ponds can also be used to collect mitigated stormwater, thereby decreasing the amount of runoff.

61. The answers are: (B) building compliant with all LEED CI credits involving stormwater design, heat island effect, and light pollution reduction
(C) building that earned 40 points under the LEED EBO&M rating system

Locating the tenant space in a LEED-certified building earns a LEED CI project team five points. 40 points is the minimum number of points a project team can earn to achieve LEED certification under the EBO&M rating system. Locating a tenant space within a LEED-registered building is a good choice; however, registration does not guarantee certification, and the building must be LEED certified prior to submission of the LEED CI documentation for five points to be earned.

For tenant spaces not located in LEED-certified buildings, SS Credit 1, Site Selection, is worth one to five points. Fulfilling the requirements of rate, quantity, and treatment of stormwater, as outlined in Option 2, Paths 2 and 3, will earn the team two points. Fulfilling the requirements for both heat island effect options under SS Credit 1 will also earn the team two more points. If the requirements are met for Option 2, Path 6, Light Pollution Reduction, the project is eligible to earn one additional point.

Energy performance, recycled content, certified wood, and green power are not covered under SS Credit 1, Sustainable Sites.

62. *The answers are:* (B) exposure through GBCI website

   (D) third-party recognition as environmentally responsible

GBCI is a third-party nonprofit organization that recognizes sustainable buildings through evaluation and certification under the LEED rating systems. Upon certification under the LEED CI rating system, the project name is uploaded to the list of LEED projects on the GBCI website.

The LEED rating systems give a snapshot of a building's performance. It is the tenant's responsibility to continue the mechanical system maintenance and minimize energy use throughout its life cycle.

Use of USGBC logos (including the LEED logo) and USGBC discounts are available only to USGBC company members.

63. *The answers are:* (D) provide ventilation rates that meet the requirements of ANSI/ASHRAE 62.1-2007

   (E) reduce water use to a level at least 20% less than the requirements of the Energy Policy Act of 1992

WE Prerequisite 1, Water Use Reduction, requires project teams to address the water usage of the plumbing fixtures. EQ Prerequisite 1, Minimum IAQ Performance, addresses the project's required ventilation rates.

Designating a commissioning authority to perform at least one commissioning design review of the owner's project requirements and basis of design is a requirement of EA Credit 2, Enhanced Commissioning. Installing HVAC equipment in compliance with ASHRAE 55-2004 is a requirement of EQ Credit 7.1, Thermal Comfort—Design. Recycling mercury-containing lightbulbs is not addressed by any of the prerequisites or credits of the CI rating system.

64. *The answers are:* (A) demonstrate a quantifiable reduction of automobile usage

   (E) verify that home offices have been set up

An ID point may be achieved if a quantifiable environmentally friendly policy or design is implemented that goes beyond the scope of the LEED CI rating system. Since the thresholds are not predetermined, it would be wise for a LEED project to submit Credit Interpretation Requests to confirm eligibility for ID credit. The TAGs that provide the credit interpretation rulings are not intended to serve as design consultants, so any Credit Interpretation Request submitted should be explicit in the project team's plan for implementation.

A home office must include all items within a person's actual office so that business can be conducted as if employees were actually at the office. Necessary items may include a laptop, cell phone, internet access, email access, printer, fax machine, reference books, paper, and pencils.

Project teams can earn two points for SS Credit 3.2, Alternative Transportation—Bicycle Storage and Changing Rooms, by providing secure bicycle racks and/or storage within 200 yards of the main building entrance for 5% or more of tenant occupants and provide shower and changing facilities in the building, or within 200 yards of a main building entrance, for 0.5% of full-time equivalent (FTE) occupants.

Project teams can earn six points for SS Credit 3.1, Alternative Transportation—Public Transportation Access, by offering a shuttle service that picks up employees within a quarter mile of the building's main entrance and takes them to the nearest bus station.

65. *The answer is:* **(A)** achievement of one ID point

Implementing the requirements of a credit from a different LEED rating system may earn a project team a maximum of one ID point, as with any innovative policy or design. Every prerequisite must always be achieved for project certification.

66. *The answer is:* **(D)** the project may not file for recertification

Recertification is only available under the LEED EBO&M rating system. There is no opportunity for recertification under the LEED CI rating system.

67. *The answers are:* **(B)** meet the requirements of the Carpet and Rug Institute, Green Label Plus testing program
(D) use adhesives with a VOC limit less than of 50 grams per liter

To achieve EQ Credit 4.3, Low-Emitting Materials—Flooring Systems, the project team must meet the requirements of the Carpet and Rug Institute's Green Label testing program, and use carpet adhesives with VOC limits less than 50 grams per liter.

Purchasing the carpet from a local manufacturer may help achieve a point under MR Credit 5, Regional Materials. Hard surface flooring may achieve a credit in the ID category. The requirements of Green Seal GC-3 are referenced in EQ Credit 4.2, Low-Emitting Materials—Paints and Coatings, not in EQ Credit 4.3. There is no requirement regarding the ratio of hard surface floors to carpeting for EQ Credit 4.3.

68. *The answer is:* **(B)** 30%, 1 point

ANSI/ASHRAE 62.1-2007 addresses the required ventilation rates determined to be acceptable for indoor air quality. Providing ventilation according to this standard is a prerequisite of the LEED CI rating system. Exceeding the required ventilation rates of ANSI/ASHRAE 62.1-2007 by 30% in all of the occupied spaces will earn one point for EQ Credit 2, Increased Ventilation.

69. *The answers are:* **(B)** divert demolition debris prior to occupancy
(E) provide a written statement that no building retrofits or remodels were performed during the certification process

Two points are available under MR Credit 2, Construction Waste Management. One point can be achieved if the landlord diverts demolition debris prior to occupancy, or if a written statement is provided stating that no building retrofits or remodels were performed during the certification process.

Work performed after building certification need not meet the requirements of this credit, though it is good environmental design practice to do so. Because there is no guarantee of what a monetary donation will be spent on, only donations of material to nonprofit organizations may qualify as diversion from landfill. Meeting a LEED NC prerequisite will not contribute toward credit earning under the LEED CI rating system.

70. *The answers are:* (A) achieve one point toward LEED certification for LEED-registered projects
    (B) be listed on the GBCI's directory of LEED APs
    (C) demonstrate a thorough understanding of the LEED CI rating system and LEED process

Achieving a satisfactory score on the LEED AP test indicates that the LEED AP has demonstrated knowledge of the LEED rating system. Including a LEED AP as a principal participant on a LEED registered project can earn one point towards LEED certification. LEED APs can choose to have their contact information listed on GBCI's website.

Companies that are members of USGBC may receive discounts on LEED reference guides and project registration fees. However, these discounts are not extended to LEED APs without USGBC membership.

All principal project team members have access to previously posted credit interpretation rulings once their project has been registered for LEED certification.

71. *The answer is:* (A) exceed the minimum HVAC system performance requirements of ANSI/ASHRAE/IESNA 90.1-2007 by 30%

The two components of EA Credit 1.3, Optimize Energy Performance—HVAC, Option 1, are worth five points each and regard equipment efficiency (C) and appropriate zoning and controls (D). Both components must be implemented to achieve ten points.

A project team following Option 2 of EA Credit 1.3 will earn five points for exceeding the minimum HVAC system performance requirements of ANSI/ASHRAE/IESNA 90.1-2007 by 15%, and ten points for exceeding them by 30%.

Installing Energy Star-rated equipment for 90% (by rated power) of the space's eligible equipment will earn the project team four points under EA Credit 1.4, Optimize Energy Performance—Equipment and Appliances.

72. *The answer is:* (B) 50%

One way to comply with SS Credit 1, Option 2, Path 4, Heat Island Effect—Non-Roof, is for 50% of a parking lot area to have a net impervious area of less than 50%. It is also possible to comply through shading, minimum SRI, parking lot covering, and/or using a building that complies with the NC rating system's Heat Island Effect—Non-Roof credit.

An impervious surface is a surface that promotes the runoff of precipitation instead of infiltration into the subsurface. Perviousness is a surface's allowance of moisture to soak into the earth below the paving system. Third-party documentation of the installed paving system's perviousness is required for compliance with the credit.

73. The answer is: (D) EQ Credit 7.2, Thermal Comfort—Verification

EQ Credit 7.2, Thermal Comfort—Verification, requires that the project team provide a thermal comfort survey to occupants, to be implemented within a period of 6 to 18 months after occupancy. The team must also agree to develop a plan for corrective action in accordance with ASHRAE 55-2004 and EQ Credit 7.1, Thermal Comfort—Design, if more than 20% of occupants are dissatisfied with thermal comfort in the building.

EQ Credit 6.2, Controllability of Systems—Thermal Comfort, provides the occupants with thermal comfort system control and does not address the survey required in EQ Credit 7.2. EQ Credit 2, Increased Ventilation, provides additional ventilation to improve indoor air quality and does not directly relate to thermal comfort.

74. The answer is: (A) bicycle storage requirements

The number of bicycle spaces needed for SS Credit 3.2, Alternative Transportation—Bicycle Storage and Changing Rooms, is calculated using the full-time equivalent (FTE) calculation, which includes transient occupancy.

The FTE calculation is not used for calculating HVAC zoning, water use reduction, or volatile organic compound (VOC) limits.

75. The answer is: (B) 10 cu ft per minute

If the ventilation rates required by ANSI/ASHRAE 62.1-2007 cannot be achieved, the mechanical contractor must test and balance the HVAC systems to ensure that 10 cu ft per minute of outside air is provided for every occupant during the occupied cycle. If 10 cu ft per minute of outside air cannot be provided for every occupant using the existing equipment, the system must be replaced, or additional ventilation equipment must be installed.

If a major renovation to the existing HVAC system is needed, the building may better qualify for evaluation under the LEED NC rating system. A renovation is considered major if more than 50% of the system is replaced.

76. The answers are: (A) drawings and photos of the project
(B) LEED CI project scorecard
(D) overall project narrative

The registration application—including registration fee, project narrative, drawings, photos, and project scorecard—establishes a record of the project. The review team can then have easy access to the principal elements of the project. After registration, the project team gains access to the online record keeping and submission tools, as well as the library of previously reviewed CIRs and rulings. The LEED CI rating system does not require a LEED AP to be a principal participant on the design team, so the certificate is also not required at registration.

No product is identified as a "LEED product."

77. The answers are: (A) building automation systems
(C) HVAC systems
(E) lighting systems and safety systems

Fundamental building commissioning verifies that a building's systems were installed and are operating as the designer had originally intended. Mechanical systems that are operated outside of the design parameters can waste energy and cost the owner more money in energy bills. EA Prerequisite 1, Fundamental Commissioning of Building Energy Systems, requires commissioning of mechanical and passive HVAC & R systems and associated controls, lighting controls (including daylighting), domestic hot water systems, and renewable energy systems.

EA Credit 2, Enhanced Commissioning, addresses future checks of the system's operations and requires the commissioning authority to develop a long-term plan for training and maintenance.

78. The answer is: (D) ASTM E1903-97, Phase II Environmental Site Assessment

ASTM E1903-97, Phase II Environmental Site Assessment, referenced in SS Credit 1, Option 2, Path 1, Brownfield Redevelopment, addresses the potential presence of a range of contaminants.

79. The answer is: (C) plumbing fixture water usage

The Energy Policy Act of 1992 sets efficiency levels for plumbing fixture water usage performance. The types of plumbing fixtures that must be considered to establish a baseline include faucets, showerheads, water closets, and urinals. Water used in drinking fountains is not included in the building's baseline water usage. Water usage of process equipment, clothes washers, and dishwashers are not included in the baseline calculation, but water reduction for these items may contribute toward ID credits.

The allowable limits of the Energy Policy Act of 1992 are 1.6 gallons per flush for water closets, 1.0 gallons per flush for urinals, 2.5 gallons per minute for faucets, and 2.5 gallons per minute for showerheads.

80. The answers are: (B) LEED Online submittal templates
(C) rulings on previously submitted Credit Interpretation Requests
(D) project access code

Registering for LEED certification starts the certification process. After registration, the LEED team leader receives the project access code, which is shared among project team members to allow collaborative work from different locations. Only registered project team members can view previously submitted Credit Interpretation Requests and the rulings on them. Registered teams upload all project information using the online submittal templates.

A list of LEED APs is available to the public on GBCI's website. A LEED-registered project can only get discounts on LEED reference guides if it is a USGBC company member.

81. The answer is: (C) Regional Priority

When a project earns a credit that USGBC has identified as being of particular importance within the project's environmental zone, that project will achieve an additional point towards project certification under the Regional Priority category.

The Innovation in Design category awards points for greatly exceeding the requirements of existing LEED credits, implementing a green strategy not addressed by the rating system, or including a LEED AP on the project team. The Indoor Environmental Quality category awards points for strategies that help optimize the indoor environment. The Sustainable Sites category addresses the project's interaction with the local ecology.

82. *The answer is:* (D) tertiary

Tertiary treatment is the highest form of wastewater treatment. It includes the removal of organics, solids, and nutrients as well as biological or chemical polishing.

Potable water meets the EPA's drinking water quality standards. Graywater is untreated wastewater that has not been in contact with toilet waste. Blackwater is wastewater discharged from toilets and urinals.

83. *The answers are:* (A) manufacturer
(B) distributor

A company that produces, stores, sells, promotes, or trades forest products is awarded chain of custody certification after auditing verifies proper accounting of material flows and proper use of the Forest Stewardship Council name and logo. LEED credits don't require that installing contractors and end users be audited in this way.

84. *The answers are:* (B) open
(D) collaborative

In traditional project delivery, teams are assembled on a "just as needed" basis and are strongly hierarchical and controlled. Integrated project teams are built to promote an open and collaborative process that optimizes project results through all phases of design, fabrication, and construction.

85. *The answers are:* (A) be actively instructional
(C) include two initiatives

Actively instructional green educational programs that include two initiatives (comprehensive signage program, case studies, guided tours, website, and so on) will achieve an Innovation in Design credit. To earn the credit, the program need not be provided by a LEED faculty member or be approved through the USGBC Education Provider Program.

86. *The answer is:* (C) ID Credit 1, Path 2, Exemplary Performance

Project teams that address the water consumption of dishwashers and laundry machines may earn an exemplary performance point within the Innovation in Design category. Regional Priority points are not awarded for exemplary performance strategies.

87. *The answer is:* (C) 10 years

Long-term lease agreements help reduce the negative impact on the environment caused by tenant relocation. Long-term agreements also require the tenant to thoroughly evaluate effects on energy and the environment before deciding to occupy the space.

Practice Exam Part Two Solutions

88. *The answers are:* (A) users are not limited to local sources of RECs
(C) RECs do not need to be delivered to the user
(E) a supplier of RECs doesn't need to meet the user's real-time electricity needs

A user can buy the least expensive REC available anywhere, not just nearby. There are no transmission costs for RECs as there are for Green-e power. Buying an REC pays for a quantity of green energy that has been added to the electricity grid, but it is separate from the electricity itself.

The energy associated with RECs and Green-e power must meet the same requirements to be considered renewable energy. As the demand for Green-e power increases, utility companies will typically create more renewable energy sources to meet the demand rather than increase the price.

89. *The answers are:* (B) conference room
(C) classroom
(D) reception area

Regularly occupied spaces are areas in which workers are located throughout the work shift or in which individuals attend a class. All areas in residential buildings are considered regularly occupied except bathrooms, utility rooms, and storage areas.

90. *The answer is:* (C) HFC-134a

CFC-based refrigerants contain the highest ozone depletion potential. HCFC-based refrigerants have an ozone depletion potential between 0.1 and 0.5. The ozone depletion potential of HFC-based refrigerants is considered to be zero. (Natural refrigerants such as carbon dioxide, ammonia, and propane also have zero ozone depletion potential.)

91. *The answers are:* (A) on-site retention ponds
(B) infiltration basins

Retention ponds capture stormwater runoff and remove pollutants prior to release. Infiltration basins promote subsurface infiltration of stormwater runoff through temporary surface storage.

Impervious surfaces increase the stormwater discharge rate. Building graywater is independent of the stormwater runoff rate.

92. *The answer is:* (B) 10%

ASHRAE 90.1 is the building energy performance standard. The Green Interior Design & Construction reference guide requires that all lighting power densities exceed the requirements of ASHRAE 90.1 by 10%.

93. *The answers are:* **(B)** design
(D) construction

The LEED AP certificate can be submitted in either the design or review application. The project team must demonstrate that the LEED AP had involvement in both phases of the project to receive the LEED credit.

94. *The answers are:* **(A)** installing a control system that provides an unoccupied mode
(C) connecting lights to motion sensors

Building automation systems can reduce energy use by optimally controlling HVAC systems. Lights controlled by motion sensors require electricity only when people are present.

Faucets with motion sensors reduce a building's use of potable water; however, water is not considered an energy commodity. Setting the building's temperature to low during unoccupied times may actually increase energy consumption due to longer morning warm-up cycles. Greater ventilation rates can improve indoor air quality, but then more energy may be needed for heating or cooling.

95. *The answer is:* **(C)** densely occupied spaces

Areas with an occupant density of 25 people or more per 1000 square feet are considered to be densely occupied. People exhale carbon dioxide. Increased levels of carbon dioxide can lead to drowsiness and lethargy.

Garages show increased levels of carbon monoxide and nitrogen dioxide that can lead to death. Smoking rooms contain thousands of different compounds, many of which cause cancer. Freshly painted rooms may experience offgassing of volatile organic compounds (VOCs) which can lead to sickness of the building occupants.

96. *The answer is:* **(A)** urea-formaldehyde

Glues used to connect composite wood and agrifiber building products may contain urea-formaldehyde.

Polychlorinated biphenyls (PCBs) were once used as stabilizing additives in flexible PVC coatings for electrical wiring and are considered to be organic pollutants. Chlorofluorocarbons were once widely used as refrigerants; they lead to the depletion of the stratospheric ozone layer. Fly ash is a by-product of incineration and can be used as a substitute for some of the portland cement in concrete.

97. *The answers are:* **(B)** reduced evapotranspiration rate
(C) isolation of graywater from human contact

Drip irrigation delivers water to the site's landscaping by means of buried pipes. The water enters the soil through perforated tubes. This method eliminates water loss from evaporation and provides a safe path for graywater.

Water is typically an inexpensive commodity. Quick paybacks are achieved through strategic landscape design rather than efficient irrigation systems. Reducing the water flow rates and flush rates of plumbing fixtures is desirable but unrelated to the choice of irrigation system.

98. The answer is: (D) benchmark the weather-normalized energy usage of the project against similar occupancies

The Energy Star Portfolio Manager program allows a project team to benchmark its building's energy usage against similar buildings. Weather-normalization allows for variations in temperature at different project locations. The program does not include a list of energy saving strategies or provide access to the energy consumption data of other projects. A LEED-certified project does not necessarily qualify as an Energy Star building.

99. The answers are: (A) operable windows
(C) occupant-adjustable volume dampers on supply grilles
(E) thermostats

Operable windows, occupant-adjustable volume dampers on supply grilles, and thermostats all allow occupants to control their thermal environment.

Occupant-adjustable volume dampers on return grilles can lead to space pressurization, called exfiltration. Thermal comfort surveys can help identify indoor air quality, temperature, and humidity problems; however, they do not provide any control of the thermal environment.

100. The answers are: (A) The enclosed floor area of a commercial interior project must be at least 250 square feet.
(B) The project's energy usage data must be made available to GBCI and USGBC for five years after certification.
(E) All applicable environmental laws and regulations must be followed.

There are seven minimum program requirements, or MPRs, that all LEED projects must meet. GBCI reserves the right to revoke LEED certification on discovery that these requirements are not met.

There is no requirement that buildings must be recertified. There is an MPR concerning the ratio of building area to site area, but it is that the building area must be at least 2% of the site area.